D1187707

SEA FLOOR SPREADING AND
CONTINENTAL DRIFT

GEOPHYSICS AND ASTROPHYSICS MONOGRAPHS

AN INTERNATIONAL SERIES OF FUNDAMENTAL TEXTBOOKS

Editor

B. M. McCormac, *Lockheed Palo Alto Research Laboratory, Palo Alto, Calif., U.S.A.*

Editorial Board

R. Grant Athay, *High Altitude Observatory, Boulder, Colo., U.S.A.*

P. J. Coleman, Jr., *University of California, Los Angeles, Calif., U.S.A.*

D. M. Hunten, *Kitt Peak National Observatory, Tucson, Ariz., U.S.A.*

J. Kleczek, *Czechoslovak Academy of Sciences, Ondřejov, Czechoslovakia*

R. Lüst, *Institut für Extraterrestrische Physik, Garching-München, F.R.G.*

R. E. Munn, *Meteorological Service of Canada, Toronto, Canada*

Z. Švestka, *Freiburg im Breisgau, F.R.G.*

G. Weill, *Institut d'Astrophysique, Paris, France*

VOLUME 2

SEA FLOOR SPREADING AND CONTINENTAL DRIFT

by

JEAN COULOMB

D. REIDEL PUBLISHING COMPANY

DORDRECHT-HOLLAND

L'EXPANSION DES FONDS OCÉANIQUES ET
LA DÉRIVE DES CONTINENTS

First published by Presses Universitaires de France, Paris, 1969
Translated from the French by R. W. Tanner

Library of Congress Catalog Card Number 79–179891

ISBN 90 277 0232 2

29 OCT 1973

All Rights Reserved
Copyright © 1972 by D. Reidel Publishing Company, Dordrecht-Holland
No part of this book may be reproduced in any form, by print, photoprint, microfilm,
or any other means, without written permission from the publisher

Printed in The Netherlands

73   02375

FOREWORD

Studies of the magnetic anomalies paralleling ocean ridges have furnished partisans of continental drift with decisive arguments. To take stock of this important question, my colleague Thellier and I decided in the early summer of 1967 to make it the subject of the annual seminar on Earth physics for the school year 1967–68. Although research was still developing rapidly, the General Assembly of the International Union of Geodesy and Geophysics held in Switzerland in September, particularly some of the meetings in Zurich under the auspices of the International Committee for the Upper Mantle, appeared to confirm that we had made no important omissions. At the conclusion of the seminar, where I had been responsible for most of the lectures, I resolved to write the present volume for the non-specialized scientific reader.

The project turned out to be a good deal more ambitious than I had thought. It is quite an undertaking nowadays to try to survey a rapidly growing subject, first of all because of the difficulty of gathering material; publication delays are now nearing one year, with the result that specialists communicate largely through a selective distribution of reports, as well as verbally in frequent colloquia. I warmly thank those who helped me in getting unpublished literature, especially Xavier Le Pichon. Even so, some essential work came to my knowledge only lately.

Finally, from the viewpoint of this Series, the subject here dealt with is certainly Living Science, and seems ripe for exposition without too much fear that premature interpretations may be rapidly superseded, for the probable lines of development are fairly clear.

It may appear that not much would be needed to make the book into an introduction to original research by adding to the already long bibliography, giving the principles behind some of the calculations, and tempering each assertion with a statement on the conditions of its validity. But in fact this would have taken me far afield to satisfy a restricted public without perhaps answering the particular questions which always occur to enquirers from another discipline. For despite the number of general surveys published every year, scientists who can use it still prefer the time-honored method of consulting a university colleague or quizzing a neighbor at a colloquium when they need information on some particular aspect not in their field.

The book has been kept relatively small then, for these reasons, and I hope that its readers will find as much interest in the wonderful recent discoveries about the ocean floor as I did in coming to know them fully.

INTRODUCTION

The author of a scientific account is always tempted by the inductive procedure; facts first, followed by interpretation. But he quickly finds this approach slow and cumbersome and soon he is putting out hypotheses. This will be my course, and I reluctantly give up, as well, attempts to follow historical development or to indicate systematically priorities of discovery, with all the difficulties attendant on the preprint system. Specialized bibliographies may be consulted (Fox, 1967).

Let me mention, however, four lines of approach to present knowledge:

(1) Location of seismic epicenters and surveys of the ridges by echo sounders, from which comprehensive results have been available since 1959;

(2) Seismic refraction, by whose aid Maurice Ewing showed in 1959 the presence of anomalous mantle under the Atlantic ridge, and Menard in 1960 the absence of crustal thickening under the East Pacific ridge; the complementary gravimetry (from 1948 to 1965, roughly);

(3) From 1952 onwards, heat flow studies initiated by Bullard; first results connected with the ridges date from 1959;

(4) Finally, and most important, magnetic studies, facilitated by the invention of the proton magnetometer (1954) which led to the discovery of the first aligned anomalies and the great transverse fractures by Mason and Raff in 1961, then, thanks to the ideas of Hess and Dietz, to the theory of Vine and Matthews in 1963, starting point of a new era.

From this geophysical break-through the subject has spread out explosively into neighboring disciplines, especially geology. It would be hard to pass over petrography and stratigraphy in this account, but conscious of my shortcomings, I say as little as possible about them.

Two notes in conclusion; year is abbreviated y. and million years m.y. Durand (1965) or (Durand, 1965) refers to the references, but mention of what Durand wrote in 1965 does not.

Addendum to the English version: I express my warmest thanks to Mr. Tanner for his excellent translation.

TABLE OF CONTENTS

CHAPTER I

SEISMIC GEOGRAPHY AND OCEAN BATHYMETRY

Almost all earthquakes are tectonic; they originate as a fracture at a point in the Earth's interior called the *focus;* the nearest surface point is the *epicenter*. The break is then propagated along the surface of a fault. Determinations of epicenters and depths of focus are carried out regularly in various institutions, such as the Bureau Central Sismologique of Strasbourg, using the times of arrival of the seismic waves at several hundreds of stations more or less well distributed around the world.

Until about 1930 when Wadati established the existence of deep focus earthquakes beyond a doubt, fractures were not thought to occur elsewhere than in the crust. In fact deep focus earthquakes are not very frequent, constituting less than 10% of the total. Their number falls off rapidly down to 300 or 350 km, this depth may be taken as a limit between the 'intermediate' and really deep earthquakes. Earthquakes become rare around 600 km and cease completely around 720 km. But these are averages; the distribution varies greatly from region to region.

In Figure 1 the epicenters of 'normal' focus earthquakes, that is, not deep focus, are plotted on an equal-area projection, using only larger magnitude events. They are seen to fall almost exclusively into two kinds of seismic zones: on one hand regions of Tertiary folding, on the other a group of elongated submarine shallows sometimes continued into the heart of the continents by regions of fracture and collapse. Epicenters are rare in the central Pacific and in the continental shields, which are stable regions surrounded by the zones previously mentioned. Earthquakes corresponding to ancient folding may be numerous (Scotland), but they are very weak.

1. Tertiary Folding, Transcurrent Faults, Island Arcs

The Tertiary foldings fall mainly into two great zones; the Pacific Belt (open at the south from Patagonia to Macquarie Island) contains four-fifths of the known epicenters. The Mediterranean or Alpine belt bordering Eurasia on the south branches from the first around the Moluccas and ends at Gibraltar.

These two belts are characterized by numerous arcuate structures; the island arcs of the western Pacific or eastern Mediterranean, the arcs of the Antilles or the South Sandwich Islands (Figure 2), the Himalayan, Carpathian ranges and so on. Most of these arcs outline closed areas: epicontinental seas such as the Sea of Japan, or sedimentary basins such as the Pannonian; but the Bonins and Marianas, or the South Sandwich Islands apparently have one free end, and the structure is very complex in the region of the Melanesian arcs.

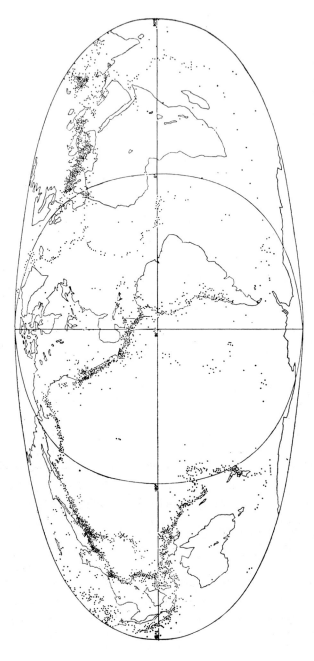

Fig. 1.   Geographical distribution of the epicenters of normal earthquakes 1930–1941. (From International Seismological Summary determinations. Equal area projection Eckert and Thomas, *La Terre*, Gallimard "Encyclopédie de la Pléiade", 1959.)

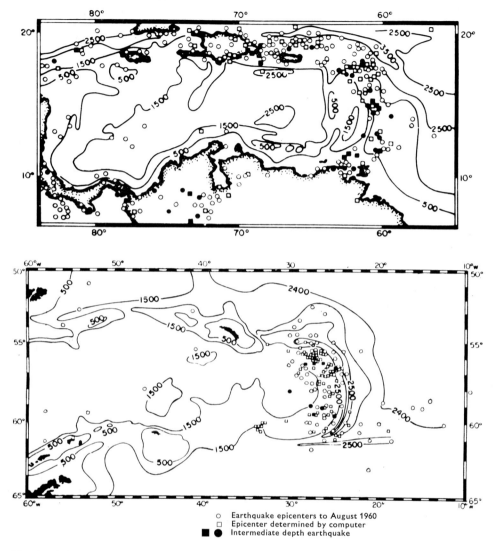

Fig. 2. Seismicity of the Antilles and South Sandwich arcs. (Heezen and Johnson, The South Sandwich Trench, Deep Sea Research, 1965.)

The movements causing earthquakes in the circum-Pacific belt, and perhaps also certain quakes of the Mediterranean zone (North Anatolian scar) are often connected with very large faults or systems of faults in which the horizontal displacement or slip is much greater than the vertical displacement or throw. The most striking feature of these faults is their approximate linearity over hundreds of kilometers. The prototype is the San Andreas fault, responsible for many known earthquakes including the much-studied one which destroyed San Francisco on April 18, 1906. During the 1906 event displacements of up to several meters at some places occurred

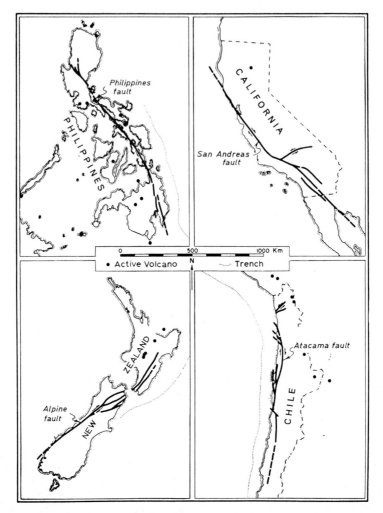

Fig. 3.   Great Circumpacific faults: Philippines, California, New Zealand, Chile. (Allen, 1965.)

on two sections of 320 and 150 km length; the fault can be followed for at least 900 km. Recent displacements, all in the same direction, are revealed by repeated triangulation; they are rarely continuous in time and generally result from large or small earthquakes; rates average several centimeters per year. From study of the terrain on opposite sides of the fault, horizontal displacements of several hundreds of kilometers since the Cretaceous are indicated (Noble; Hill and Diblee; Crowell) corresponding to a mean rate an order of magnitude smaller than this.

   Along the Pacific belt, Allen (1965) has systematically investigated large trans-current faults with visible displacement to see whether they were associated with historical earthquakes. The principal examples are given in Figure 3. Study of thalwegs deformed by the fault gives the direction of motion. The Philippine and Formosan faults are exceptions to an alleged curious rule that the motion on all faults corresponds to a counterclockwise rotation of the central Pacific.

Deep focus earthquakes are confined to regions of Tertiary folding, as far as is known, but they do not occur in all such regions, being absent for example from the North American coast of the Pacific rim. Their presence is generally associated with that of an ocean trench, that is, a long narrow steep-flanked depression (order of magnitude: over 1000 km long, 100 km wide at the top, 10 km at the bottom). Trenches are generally asymmetric, the slope being steeper on the side bordering the island arc or continent. The bottom is sometimes flat, but this seems to be due to filling by turbidity currents. The trenches accompanying the island arcs in the western Pacific, where the ocean is already very deep, contain the greatest known depths, often greater

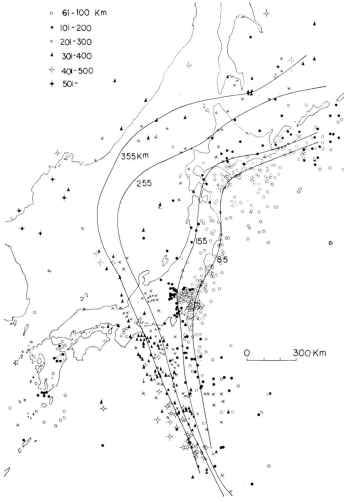

Fig. 4. Epicenters of deep-focus earthquakes from 1926 to 1956 determined by the Japanese Meteorological Service. (Contours from Sugimura and Uyeda, *Japanese national report on the upper mantle*, 1967.)

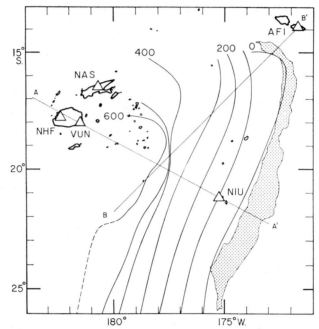

Fig. 5. Deep-focus surface in the Tonga trench region. Triangles denote the Lamont Geological Observatory temporary stations which were used to locate the foci: NIU, Tonga; AFI, Samoa; NAS, NHF, Fiji. Stippled area: ocean depth over 6 km. Contour interval 100 km. (Oliver and Isacks, 1967.)

than 10 km, while those of the Mexican coast rarely exceed 5 km. All are readily distinguishable from the great aseismic depths of the North Pacific.

Except for the region of the New Hebrides and the Solomon Islands, or the re-entrant angle near the boundary of Chile and Peru, the trench is concave toward the neighboring continent. In this direction, about 100 km distant, occurs a chain of volcanoes producing acid lavas. Focal depth increases fairly regularly; normal focus earthquakes are near the axis of the trench but more numerous on the continental side; intermediate foci are largely under the chain, and the really deep quakes notably beyond it. The foci as a group lie thus on surfaces inclined at about fifty degrees, usually only occupied at certain depths, but which have been fairly clearly defined in at least two cases (Figures 4 and 5). Sykes points out that for the Tonga earthquakes (Figure 5) the curvature of the trench in the horizontal plane is followed down to 600 km despite the intervening perturbations.

In the Tyrrhenean Sea and the eastern Mediterranean with similar arcuate structures, intermediate depth earthquakes are found with an analogous pattern. A surprisingly deep focus at 640 km has been observed for the earthquake of 29 March 1954 in the concavity of the Betic Cordillera. Finally, in the continental regions of the South Asian rim, great arcuate structures, including the Himalayas, are not accompanied by deep earthquakes throughout their length. The observed earthquakes, all intermediate in depth, cluster where the belt changes direction, in Burma, the Hindu Kush and the Carpathian arc. Under each of the two latter regions, the quakes oddly recur in the same place and at the same depth (225 and 150 km respectively).

2. The Crests of Ridges

The lines of ocean epicenters are sparser but more clearly traceable than the lines of Tertiary foldings. Their real width probably does not exceed 20 km (Sykes). All of the

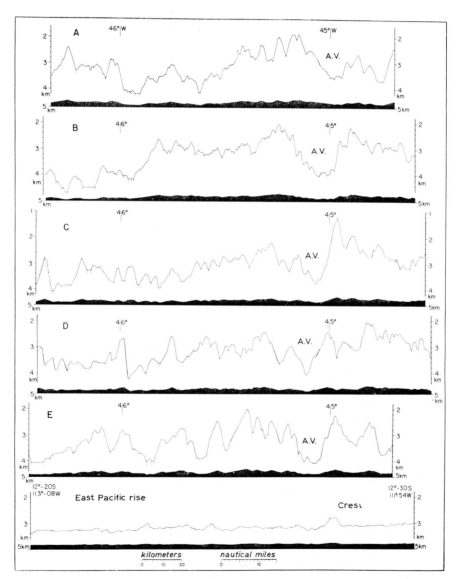

Fig. 6. Profiles of the Atlantic ridge between 22°N and 23°N, and of the East Pacific rise around 12°S. The variability of relief in the crestal region of the former is shown, particularly in the peaks and plateaus to the west of the axial valley (A.V.). (Van Andel and Bowin, 1968.)

corresponding tremors are of normal depth, and rarely very large. These lines of seismicity follow elongated shallows, rises or ridges. The typical example is the mid-Atlantic ridge discovered over a century ago in laying the first telegraph cables. The line of seismicity divides the ocean down the middle, following the curvature of its opposite coasts. The average width of the ridge exceeds 1000 km, but the earthquakes are located right on the crest (de Vanssay in 1939). The ocean depth over the crest is around 2 or 3 km, while the adjoining basins on either side are about 5 to 6 km deep. Ewing and his colleagues have found (Heezen *et al.*, 1959) that the 'crest' corresponds to a rift valley with a maximum depth of 2 km and a width of 30 km lying between two very uneven regions (Figure 6). To be exact, the median valley may be bifurcated, obstructed by volcanoes, displaced laterally, or even disappear (as for example south of Iceland on the Reykjanes ridge).

Let us start from the equator, along which the Atlantic seismic zone runs roughly from St. Paul Rocks to the Romanche trench. Going north the zone describes a large arc as far as the Azores, paralleling the African coast. From there a branch runs towards Gibraltar or Cape Saint Vincent along a poorly defined rise. Between the Azores and Iceland, another branch may connect several epicenters situated between

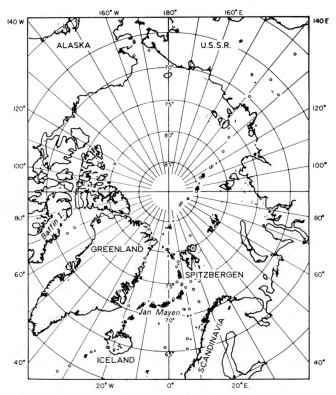

Fig. 7. Arctic epicenters, January 1955–March 1964. Circles represent epicenters determined by the author from 10 or more seismic stations, smaller circles from less than 10, crosses recent determinations by the U.S. Coast and Geodetic Survey from 10 or more stations. (Sykes, 1965.)

Labrador and Greenland, but the continuity of the two regions is not established.

Iceland is really the only island crossed by the median valley of a ridge, the others being merely adventitious volcanoes. Beyond it (Figure 7) the line passes through Jan Mayen Island, then describes a new arc passing to the west of Spitzbergen as far as the continental shelf of Greenland. It then follows the 'Gakkel Ridge', midway between Franz Josef Land and the great aseismic Lomonosov rise, turning towards the mouth of the Lena, widening as it reaches the shelf which forms the bottom of the Laptev Sea. (It is not really surprising that the oceanic seismic belts should become diffuse when they reach a continent, where faults are numerous.) As few examples are known of seismic lines stopping without evident cause, efforts have been made to link this one either with the Baikal graben, formed in the Plio-Pleistocene along earlier faults, or with the Verkhoyansk range, Mesozoic foldings with few earthquakes.

Let us return to the equator. On the south the ridge system goes to Ascension, Tristan da Cunha and Bouvet Island, where it is joined by a line coming from the South Sandwich arc, the origin being visible at the south end of the arc in Figure 2. Then the main ridge goes around both Africa and Madagascar, at first keeping midway between them and Antarctica, and then between them and the plateau of Kerguelen Island. It passes the Prince Edward Islands, reaches a junction point near Rodriguez Island, runs straight north along the 'Carlsberg Ridge' and finally enters the Gulf of Aden, where it divides on one hand in the direction of the great African lakes, on the other towards the Red Sea and the graben of the Jordan. This continuity of the central valleys of the ridges and of the African rifts has led them to be considered analogous structures, and while the morphological relationship is striking (Figure 8), we shall find profound differences.

From the Rodriguez junction another branch passes between the Kerguelen Plateau and then between Antarctica and Australia following the curvature of their shelves. It ends at about 160°W long., to the south of Macquarie Island.

The whole system just described in its seismic and bathymetric aspects may be called the Atlantic-Indian ridge or the mid-oceanic ridge to denote its remarkable property of dividing the oceans it crosses in the middle. The term mid-oceanic cannot properly be used of the following examples.

The line of epicenters does not end at 160°W, but continues, getting farther from

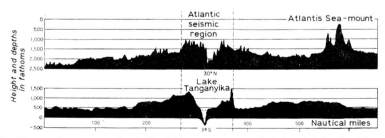

Fig. 8. Profiles on the same scale of the African plateaus and the mid-Atlantic ridge, with their median valleys. (Ewing and Heezen, 1960.)

Antarctica, until it turns northward to Easter Island. This section may be called the South Pacific ridge. The topography changes from that of the Indian ridge in that the median valley disappears (Figure 6). But the depth to the ridge remains 2 or 3 km, as for all the ridges (except where they cross continental regions, or in the fracture zones soon to be discussed).

Easter Island forms a new junction point, a seismic line reaching it from Patagonia. The main branch, or East Pacific rise, goes north, passing the Galapagos. From there to the Gulf of California, which it enters, it is linked by seismic rises to South and Central America. Two continental (and therefore diffuse) branches leave the Gulf of California. One is an inland belt of faulted regions involving the basins and ranges of Utah, the other the San Andreas system, extended all the way to Alaska by a series of coastal epicenters in which we shall later recognize fragments of ridges with central valleys.

To this general system of ridges totalling about 50000 km in length as we have described it, various authors make modifications or additions rendered plausible by the lack of precision in the lines of epicenters or in the bathymetry.

Thus Menard (1956b) divides the East Pacific rise into two by tracing an arc from Panama (possibly even from Honduras) to Patagonia (perhaps even to the Scotia Sea). On the other hand he continues the East Pacific rise through Alaska and the Canadian Arctic archipelago to meet with a branch of the Mid-Atlantic ridge passing between Greenland and North America (marked by two epicenters in Figure 7).

From Macquarie Island to New Zealand and beyond to the north northwest (Menard's Melanesian ridge) other structures sometimes assigned to the ridge system should rather be linked with the system of the island arcs.

3. Bathymetry of the Oceans

We have introduced the ocean ridges by considering their nearly linear axial region, but they are much more extensive structures. Three principal morphological divisions may be distinguished in the oceans:

(1) *continental margins* and island arcs;
(2) *oceanic basins*, comprising abyssal plains and abyssal hills;
(3) rising from these last, the *ocean ridges*.

The Atlantic ridge, for example, occupies roughly the central third of this ocean. The mean slope of its sides generally increases in approaching the crest. The location of the East Pacific ridge precludes its attachment on the east to any notable feature of the ocean floor, but Menard (1964) suggests that the ridge persists under the western part of the American continent, and we shall see how fruitful this hypothesis is.

In this chapter we confine ourselves to a general description by considering first the topography of the ridges as revealed by echo sounding, then that of the bounding surface between the more or less consolidated sediments and the usually volcanic basement, this surface being obtained by continuous reflection profiling. Other details will be added later.

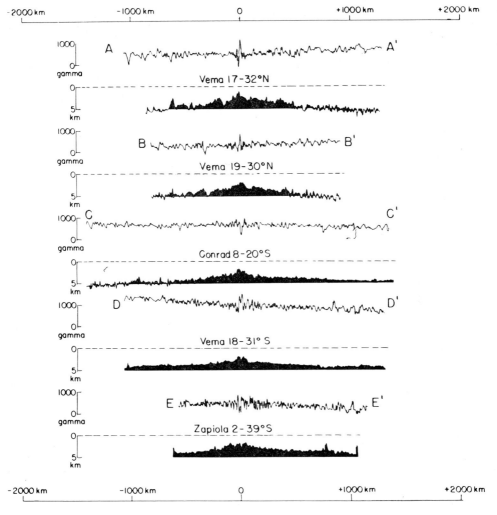

Fig. 9. Atlantic ridge traverses showing profiles of anomalies of magnetic intensity, and of the basement. Conrad 8 and Vema 18 profiles have been projected perpendicular to the crest (1 gamma = 10^{-5} Oe). Vertical exaggeration 40 times. (Heirtzler and Le Pichon, 1965.)

The mid-Atlantic ridge has been well studied by the Lamont Geological Observatory group, to whom we owe (Figure 9) five characteristic profiles of the basement surface (Heirtzler and Le Pichon, 1965). Towards the north, between Greenland and Norway, the ridge widens and becomes less deep.

To see the regional topography, wavelengths shorter than about 100 km are filtered out of the profiles, as corresponding to local irregularities too small for isostatic compensation. The crest then shows up at a uniform depth of 2.5 ± 0.5 km, but the depth over the flanks is variable. North of 29 °S, they run down a 3 or 4% slope to meet the abyssal hills at a depth of 5 to 6 km. South of that latitude the slope is very slight and the flanks form a plateau at 3.5 or 4 km depth, interrupted on the north by the Walvis and Rio Grande ridges. In both cases the basement maintains a fairly steady depth of around 6 km under the lateral basins from which the abyssal hills rise, without following the slope of the flanks. The greatest depths, on the order of 6.5 km, usually carry great thicknesses of sediment, suggesting that isostatic re-adjustment has taken place. The basement dips again to 7 km or more under the continental rise, forming a sort of trench there. Le Pichon, by taking the edge to be where the basement slopes under the flank, arrives at a width of 1000 km or so for the ridge in the northern Atlantic, and double this width south of 30°S. But there is no discontinuity between the ridge so defined and the hilly abyssal region.

The relief shown by filtering out long wavelengths reveals, throughout the length of the ridge, a band 150 to 200 km wide formed by the two blocks adjoining the central valley, free of non-consolidated sediment. There is a change in the flank profile around 29°S. To the north, the flanks are described by Heezen *et al.* as being constituted of raised and tilted blocks (distinct from the two axial blocks) of up to 100 km width, whose escarpments face the crest, like the usual displacements of graben faults. This state of affairs is not however as evident as they indicate. Still in the north, the sediments on the flanks are collected in pockets with generally horizontal surfaces. The basement is exposed on the peaks. The transition to the abyssal hills is marked by the disappearance of wavelengths exceeding 50 km. The main features of the flanks are elongated parallel to the crest; this is not as clear in the case of the abyssal hills.

South of 29°S, wavelengths over 50 km are deficient on the flanks; the relief is more localized and subdued. The roughness (filtered out) is on the same scale. The sediments, highly transparent acoustically, form an unbroken cover of some hundreds of meters thickness at most.

The difference between the flanks of the ridge north and south of 29°S attracts attention to the two aseismic rises, Walvis and Rio Grande, which depart from it in these latitudes. The crest of the Walvis rise, linear over 200 km, is at a depth of 1500 to 2000 m. Its eastern edge forms an escarpment plunging to 4500 m in a few kilometers; its western edge has a moderate slope, generally fragmented. The Rio Grande rise is a large block with a moderate slope towards the Argentine basin on the south-west, steeper towards the north and east. Le Pichon considers the two rises to be structures arising from repeated motion on Mesozoic faults.

There is a great contrast between the ridge or the rises, and the adjoining basins, notably the Argentine basin in which up to 2.5 km of continental sediments have accumulated despite the narrowness of South America at this latitude.

The morphology of the East Pacific rise (Figure 6) is less well known than that of the mid-Atlantic ridge. Its essential characteristics appear in the description of Menard (1964): symmetry about a crest lacking a median valley, much greater width than the Atlantic ridge (order of two times), less marked relief if the isolated volcanoes are disregarded, and especially that the maximum relief occurs far out on the flanks in regions of saw-toothed topography probably produced by normal faults parallel to the ridge.

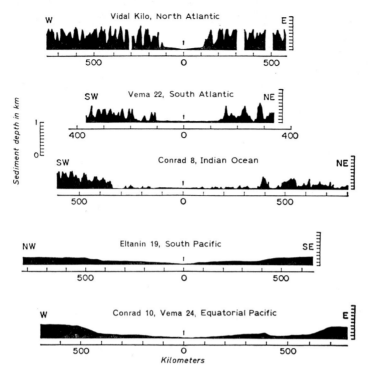

Fig. 10. Sediment thickness as a function of distance from the ridge crest. The bottom profile comes from three traverses of the East Pacific rise where it goes through the equatorial zone of thick sediments. To the east the picture is complicated by fractures or by structure associated with the Galapagos. (Ewing and Ewing, 1967.)

As for the sediments, the description given earlier for the Atlantic seems applicable to oceans in general (Ewing and Ewing, 1967), on a larger scale for the East Pacific rise. Soundings still have insufficient resolution to detect a measurable thickness on the axis. Coring and photography often reveal outcrops of the basement. The sediments grow progressively thicker on the flanks, depending on the sedimentation rate, up to 100 to 400 km from the axis; the thickness then increases abruptly and remains constant or increases only slightly on increasing distance from the axis. Figure 10 shows thickness of sediment (smoothed by averaging over 10 km) in five profiles in various oceans. The extraordinary smoothness of the Pacific profiles will be noted.

We shall not describe the continental margins and island arcs in as much detail as the ridges whose structure is fundamental for us. We shall merely recall that the margins are divided into a continental shelf off shore, a continental slope and a (precontinental) rise in the terminology of Le Pichon (1966). We shall return to the island arcs later. Although narrower than the ridges they are more extensive structures than normal focus earthquakes would suggest. To begin with, the typical trench may, as in Indonesia, be divided lengthwise by a sedimentary arc either below or above water. Thus there are simple island arcs (volcanic), and double ones (sedimentary and volcanic). In the second place, Menard (1964) associates with the arcs more or less parallel undulations which are particularly prominent when they form the boundary of an epicontinental sea. These undulations would increase the width of the Marianas arc to 1500 km.

4. The Focal Mechanism of Earthquakes and the Direction of Tectonic Stress

Two kinds of waves originate at a seismic focus. Those which arrive first, the *P* waves, correspond to to-and-fro motions in the direction of propagation, as for sound waves, while the second or *S* waves correspond to motions in a plane at right angles to this direction. In a plan view, then, the first displacements observed at stations near the epicenter will converge or diverge from it depending on the station location. Let us examine a simple case in which a vertical fault with trace *F* (Figure 11) slips hori-

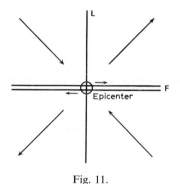

Fig. 11.

zontally; the greatest displacement is at the fault while distant points on a perpendicular line *L* will not move. It can be shown that in such a case the regions where the first motion is towards the epicenter (dilatation) are separated from the regions where the first motion is away from the epicenter (compression) by the vertical planes through *F* and *L*. Conversely, observation of compressions and dilatations in alternating quadrants locates the two planes, but does not distinguish which is which.

In the general case, after observing dilatations and compressions at distant stations it is necessary to make allowance for the complicated paths followed by the waves through the interior of the Earth before reconstructing the motion at the focus. Such work is carried out routinely for most large earthquakes. After eliminating some inconsistent observations it is possible to arrive at statistically valid conclusions on the position of the fault and its initial movement. For a dozen earthquakes due to visible faults the method has produced a satisfactory agreement with the orientation of these faults and the surface displacement.

Some fears have been expressed that the results might be of merely local significance. It is found however that at a given observatory large distant earthquakes usually give the same direction of first motion, at least when the depths are of the same order. This demonstrates both the regional character of the phenomenon and its persistency.

The theoretical difficulty in distinguishing between the fault plane and the plane perpendicular to the displacement on it concentrates attention first on the well-determined intersection of the two planes, that is to say the line along which the first motion vanishes.

The following rules seemed to emerge from the sometimes contradictory results: along the zones of Tertiary folding (excepting such regions as off-shore British Columbia, from the Solomons to the Sunda Islands, the Bonins, Central Asia), the line of zero motion is nearly vertical. Therefore both of the possible solutions correspond to predominantly horizontal movements (slip). In one case the fault-plane is more or less parallel to the tectonic belt, and in the other more or less perpendicular to it.

Slip parallel to the structure seemed to be called for in the region of the great circum-Pacific faults (San Andreas, Chile, etc.). Movements perpendicular to the coasts, which seemed necessary to explain orogenic features in these regions, would

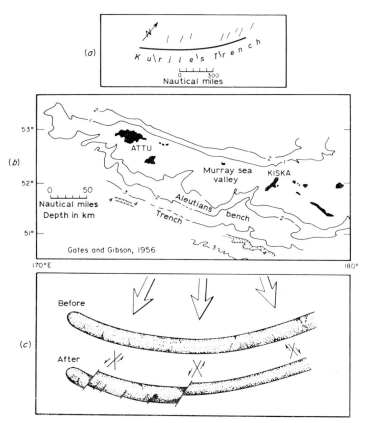

Fig. 12. (a) Oblique lineations on the slopes of the Kuriles trench; (b) Indentations and offsets on the north slope of the Aleutians trench; (c) Interpretation in terms of transverse faults. (Menard, 1964.)

correspond to other faults farther inland and slipping less frequently (the Owens Valley fault in California, responsible for the large earthquake of 1872, will serve as example).

The second solution appeared preferable for the island arcs of the western Pacific. Each arc could then be thought of as more or less fixed at its ends and pushed from inside so as to shear it in places. Observation of transverse tectonic lines in the Kuriles and Aleutians led Menard (1964) to similar conclusions (Figure 12).

Still speaking of Tertiary foldings, most of those cases which do not correspond to these two solutions in which vertical motion predominates, may correspond to motion on thrust faults, tangent to the surface on which deep foci lie. The continent then appears to override the Pacific Ocean, except in the Solomons and Bismarck Island region where the seismic surface slopes the other way and the ocean seems to override the Coral Sea, and also excepting the New Hebrides region, where the Fiji plateau acts as continent.

Coming back now to the ridges, the presence of a central rift in the mid-Atlantic-Indian Ocean ridge had early led to the belief that they were under tension and that their earthquakes were the result of normal faulting, as in the case of tremors along continental grabens.

These results, seemingly well established by 1960, gave credit to the simple idea that this two-fold division of the great seismic belts reflected a division into regions of compression or slip (Tertiary foldings) and regions of tension (the ridges).

Things seem less simple today. Sykes, who has recently carried out several detailed studies to be discussed later, believes that the poor distribution of stations and the indifferent quality of the older data have exaggerated the importance of strike slip. Moreover the nature of the stresses involved in Tertiary arcs and in their accompanying trenches is once again in question. The idea that both are due to compression seemed well founded after the famous gravimetric campaigns of Vening Meinesz and the theories he used in interpreting his results. Thrust accounted fairly well for double arcs (a sedimentary arc along the trench with a volcanic arc in the concavity) by a compression of the sediments. But we shall see that Worzel has reinterpreted the gravimetry of trenches as analogs of continental grabens. In support Menard (1964) notes in certain trenches the presence of an axial ravine cutting their floor, and in most the presence of lateral terraces forming a trap for sediments, two circumstances suggesting rather tensile stresses, perhaps accompanied by vertical ones. All of this is hard to reconcile with the foldings observed in Indonesia on the islands of the sedimentary arc. We shall return to these problems in Chapter VI.

CHAPTER II

RESULTS OF MAGNETIC MEASURES AT SEA.
TRANSVERSE FRACTURES

1. The Magnetic Anomalies of the Ridges

Measurements of the total intensity of the magnetic field made by ship-towed magneto-meters have permitted the detection of anomalies sometimes exceeding 3000 γ (1 $\gamma = 10^{-5}$ Oe) on the ridges, while values one-tenth as large are rare on the con-tinents. First results came from the East Pacific where Mason in 1958, then Mason and Raff in 1961 revealed curious parallel anomalies (Figure 13), and concluded

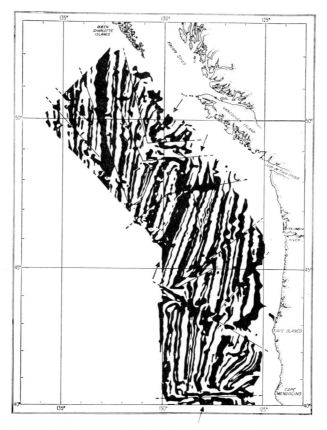

Fig. 13.   Magnetic anomalies discovered by Raff and Mason in 1961; positive anomalies in black. Faults are indicated by straight lines. Arrows are axes of three ridge segments, from north to south: Explorer, Juan de Fuca, Gorda ridges. (Vine, 1966.)

from the sharpness of their edges that they originated in magnetic masses at not over 3 km depth. We begin however as before by considering the mid-Atlantic ridge in its regions with a median valley, that is, almost throughout its length.

The corresponding magnetic anomalies are highly variable from place to place. The most intense anomalies which first attracted study are on the crest. This is clearly seen in Figure 9 of Chapter I. In this, as in all succeeding illustrations, the anomaly is the residual after removal of a global field represented by development in spherical harmonics. The great axial anomaly is accompanied by weaker anomalies closer together; these are more distinct north of 29°S than in the south where the crest zone is much wider and the axial anomaly appears multiple. The maximum anomaly follows the topographic and seismic axis very faithfully.

If the magnetic mass producing the anomaly is modeled as a uniform prism striking north and south, 10 km below the ocean floor, with a susceptibility of 0.01 emu, the total intensity can be calculated from the strength of the inducing geomagnetic field, its relative direction and the depth of the water (Figure 14). This simple hypothesis

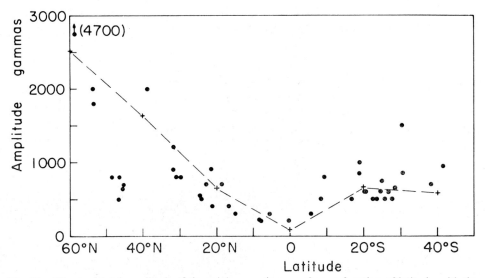

Fig. 14.   Observed total amplitude of the axial magnetic anomaly as a function of latitude, with the calculated variation for a model with constant volume and susceptibility. (Heirtzler and Le Pichon, 1965.)

is found to explain the observations well. The largest deviations occur at the transverse fracture zones to be discussed later. It appears, then, that the axial anomaly is due to the presence of a highly magnetic intrusion in the volcanic basement, directly below the median valley.

On the flanks of the ridge the wave-length of the anomalies increases, as do their amplitudes when corrected to a fixed height above the floor, 3 km for example, as in Figure 15, which shows profiles characteristic of the northern ridge. The wavelengths

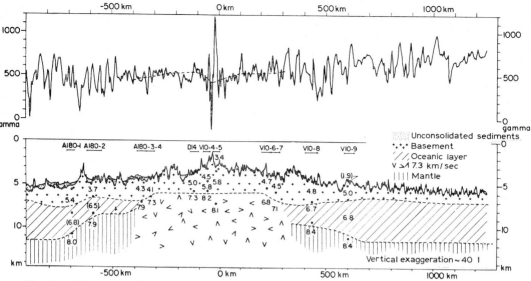

Fig. 15. Magnetic profile of the mid-Atlantic ridge at 32°N as it would appear at a constant 3 km above the bottom. *Dashed*, assumed effect of variation in the depth to the Curie point. *Below*, composite section of the crust in the same region. (Heirtzler and Le Pichon, 1965.)

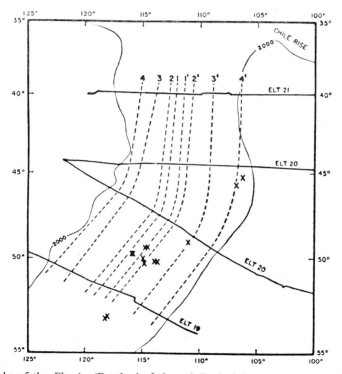

Fig. 16. Tracks of the *Eltanin*. (Depths in fathoms.) Dashed lines connect recognized magnetic anomalies of the four profiles. Crosses give approximate locations of seismic epicenters. (Pitman and Heirtzler, 1966.)

over 60 km seen on the flanks cannot be explained by the topography. The axial anomaly (1200 γ) is conspicuous.

Let us consider this axial anomaly. In spite of the relatively good fit obtained on the hypothesis of induced magnetization, this was merely a convenience for calculation, and its authors were aware that thermo-remanent magnetization played a much more important role. The bottom of the prism in reality corresponds to the Curie point isotherm of the matter constituting it. We shall find that heat flow is high on the axis of the ridges, raising the isotherm; the effect of this is shown in Figure 15.

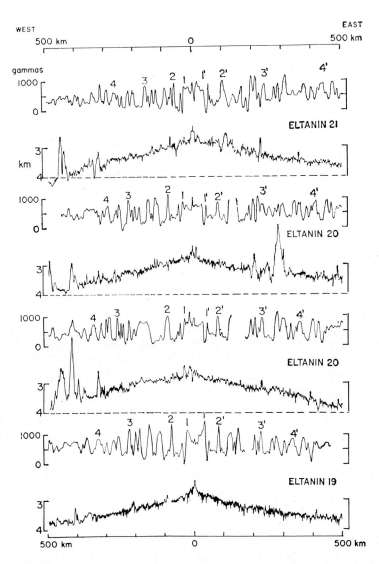

Fig. 17.   Magnetic and bathymetric profiles along the tracks of Figure 16, projected perpendicular to the ridge axis. (Pitman and Heirtzler, 1966.)

This limitation by the Curie isotherm is only one possibility however (others are presence of other rocks, changes in grain size, etc.).

Similar results were obtained in 1963 on the Carlsberg ridge (Vine and Matthews, 1963), and in 1965 (Figures 16 and 17) from four traverses of the Pacific-Antarctic ridge between 40° and 55°S (Pitman and Heirtzler, 1966). The lineation is still excellent here as shown by the agreement of the four profiles. But a new phenomenon appears: the axial anomaly 1-1′ and the lesser anomalies next it are symmetric about the axis, and this symmetry persists over quite a width, as shown by the pairs 2-2′, 3-3′, 4-4′.

Perhaps the significance of these results, suggesting the existence of longitudinal intrusions in the ridges, can best be discerned in Iceland which straddles the Atlantic ridge. The forked region of Figure 18, whose actual shape is probably even more complex and which is the seat of numerous active fissures more or less parallel, may be considered as formed from elements of the axial zone.

The oldest Icelandic rocks are found on either side of this zone towards the edges of the island. They are not very old; Gale, Moorbath, Simons and Walker in 1966 found ages by potassium-argon dating for the plutonic rocks of 1.5 to 10 m.y. (Upper

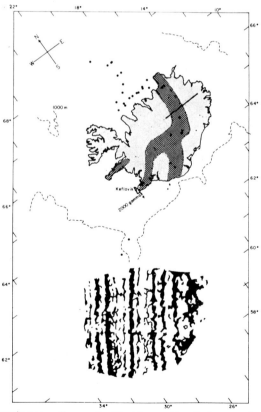

Fig. 18.   Positive magnetic anomalies on the Reykjanes ridge (epicenters from Sykes, 1965). Dark shading indicates the zone of Quaternary volcanic formations in Iceland. (Le Pichon, 1966.)

Miocene to Pleistocene) among the acid intrusions in the south-east and in the west of Iceland, and they believe the oldest basalts do not go beyond the Miocene.

Unlike those authors who see Iceland as a geosyncline or a plateau slumped in the center, Bodvarsson and Walker in 1964 considered it to be the result of a progressive outflow of material rising in the fissures of the axial zone (Walker, 1965). The visible fissures (about half of which have emitted lava, the rest remaining open) would account for a forced expansion of the basement, estimated at 30 m during the last 3000 to 5000 y. Walker has also examined the dikes of solidified lavas which intersect at right angles the benches of Tertiary lavas in the east of Iceland; a prism of lava 10 km high having the present width of Iceland would correspond to 200 to 400 km cumulated thickness for its feed dikes.

The details have been warmly discussed (Björnsson, ed., 1967), but the ideas of many current writers on the origin of the ridges and sea-floor spreading follow directly from these views of Bodvarsson and Walker on the repeated intrusion of flat juxtaposed dikes.

In such a volcanic country a land network of magnetic survey would be impractical. A recent Canadian airborne survey (Björnsson, ed., 1967) resulted in complicated profiles which became simpler over the sea south of the island.

The portion of the ridge coming off the Reykjanes peninsula had earlier been the object of an accurate aeromagnetic study covering a square of 350 km side centered on 60°N, 28°W (Heirtzler *et al.*, 1966). The ridge follows a straight course as far as 55°N. The main lines of its topography are parallel to the axis and symmetric about it; in the north the crest is bounded by sharp increases in depth at about 30 km on either side of the axis, but these contrasts diminish to the south, and at 60°N the depth increases fairly regularly with distance. Details of width under 5 km do

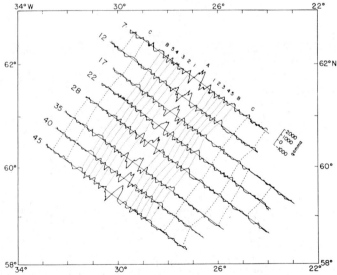

Fig. 19.   Eight profiles projected perpendicularly to the Reykjanes ridge axis. (Heirtzler *et al.*, 1966.)

not show the same parallelism. Heirtzler, Le Pichon and Baron did not find a central valley; Schilling *et al.* (1968) found one at 60° and 61°N, but less than 10 km wide, containing patches of pillow lava.

The magnetic anomalies on this Reykjanes ridge are remarkably parallel to each other (Figure 18). Magnetic intensity along the axis follows the topography faithfully (Talwani *et al.*, 1968). Figure 19 shows eight profiles perpendicular to the axis at 40 km intervals. The axial anomaly is positive, thus due to a normally magnetized body, and is large, as expected in this high magnetic latitude; the width is of the order of 40 km and the amplitude is 3000 γ. Six anomalies can be made out on either side, with an amplitude of 500 to 1000 γ and a wavelength of about 15 km. They are distributed symmetrically about the axis of the ridge. They bear little relation to the relief, whose occurrence as tilted parallel blocks hardly shows in the magnetic field.

If allowance is made for the height of observation above the bottom, the central anomaly predominates, as it does in the northern part of the Atlantic ridge; the lateral anomalies resemble more those of the southern part. But the set of crest anomalies is much more regular; only the flank anomalies more than 100 km from the axis are irregular. The axial anomaly continues more or less on the Icelandic continental shelf, but the lateral anomalies disappear.

The remarkable parallelism of the magnetic anomalies on the Reykjanes ridge recurs over the axial valley in the southern part of the Red Sea (Girdler, 1968). Figure 20 shows this axial valley and the anomalies found by Drake and Girdler near 16°N. Correlations established between the anomalies encountered by the ship along the track shown in Figure 21 enabled them to establish the plot of maxima and minima; these correlations, of course, are much less certain than in the case of the Reykjanes ridge. Girdler saw them as the result of a series of parallel intrusions

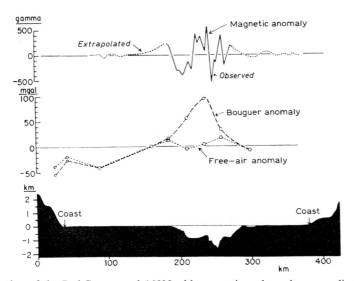

Fig. 20.   Section of the Red Sea around 16°N with magnetic and gravity anomalies. (Drake and Girdler: 1964, *Geophys. J.* **8**, 478.)

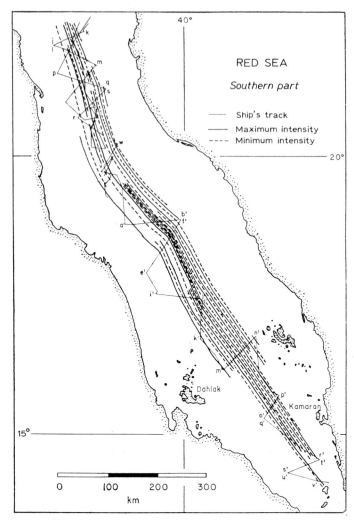

Fig. 21. Correlation of the great magnetic anomalies in the southern Red Sea. (Drake and Girdler: 1964, *Geophys. J.* **8**, 484.)

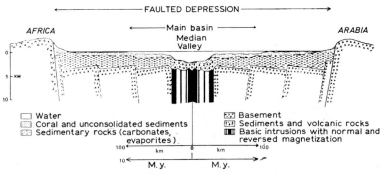

Fig. 22. Structure of the southern part of the Red Sea. (Girdler, 1968.)

separated by non-magnetized rocks in the middle of a region of extension of the crust (Figure 22).

2. Linear Anomalies in the Neighborhood of Island Arcs and Coasts

The linear character of the anomalies around ridges is found again to a certain degree in the neighborhood of island arcs as is shown in Figure 23 drawn up by Heirtzler (1965) from Japanese and Russian magnetic surveys. It is clear that here too, the matter causing these anomalies is at shallow depth. Similar anomalies run along the eastern part of the south edge of the Aleutians trench (Hayes and Heirtzler, 1968) to rejoin the anomalies of the American Pacific coast (extension of the anomalies of Mason and Raff) in the Gulf of Alaska where the system forms an acute angle.

Note in Figure 23 the presence of anomalies inside the Sea of Okhotsk, observed by Solovyev and Gainanov in 1963.

Apart from these regions the magnetic field near the trenches is little disturbed. However, as indicated in Figure 24, also compiled by Heirtzler from various sources, some less regular anomalies (except for the beginning of one along the Antilles arc) border the Atlantic coast of North America and even encroach on it. The magnetized rocks causing these marginal structures do not appear to lie farther below sea level than those of the abyssal plains, although they are buried under thick sediments. Out to sea a magnetically quiet zone some 400 km wide is found, separated by a line from the more disturbed regions which continue out to the Atlantic ridge.

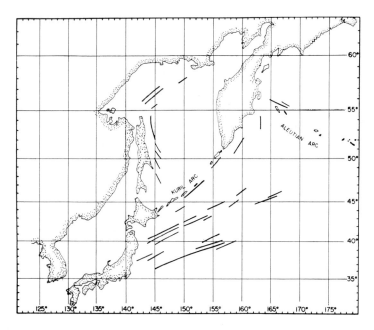

Fig. 23.   Linear anomalies in the Pacific east of Asia. Lines correspond to maximum anomalies.
(Heirtzler, 1965.)

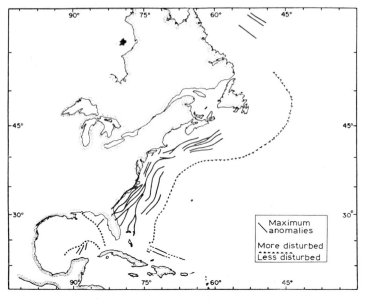

Fig. 24.   Linear anomalies in the Atlantic east of North America. (Heirtzler, 1965.)

Heirtzler and Hayes (1967) have discovered a magnetically quiet zone on the east of the Atlantic. It is narrower, yet closer to the ridge. Finally, quiet zones and some lineations have lately been discovered (see Chapter VI) near the coasts of the South Atlantic (Dickson, Pitman and Heirtzler, 1968).

3. The East Pacific Rise and the Great Transcurrent Fractures

Although the anomalies of Mason and Raff were the first oceanic anomalies discovered, we have left their discussion to this point. This is because they are perturbed by the phenomenon, discovered by Menard and Dietz in 1952 in the East Pacific where it occurs on the largest scale, of transverse fractures. These features are described by Menard (1964) as long narrow linear bands, less mountainous than the surrounding bottom, separating two regions of different depths. The great east-west fractures of the Pacific are thousands of kilometers long, 100 to 200 km wide, with often up to 3 km of relief. The network of magnetic anomalies, whose general trend is north-south, undergoes sharp discontinuities, as well as local distortions in crossing these fractures. The most northerly, the Mendocino, appears at the bottom of Figure 13. It is succeeded by the Pioneer and Murray faults, off the coast of California.

As shown by Figure 25, comparison of the anomalies on the two sides of these features allows the matching of regions displaced laterally by sometimes several hundreds of kilometers. At first this was interpreted by supposing that the magnetic layers of the sea floor had slid along the fractures without significant distortion. However, the two values of the displacement on the Murray fault, 680 km to the west, 150 km to the east would imply that between $130°$ and $139°W$ ('disturbed zone'

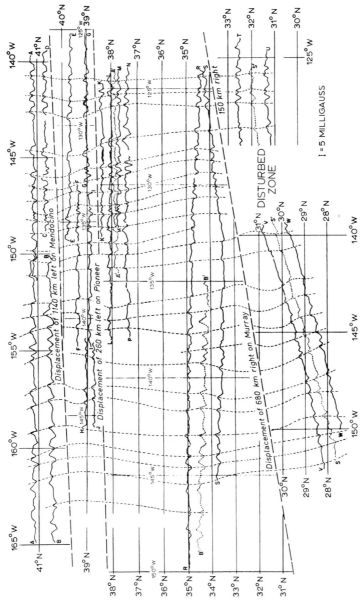

Fig. 25.   Profiles of magnetic intensities off California. The profiles labelled with primed letters, as A´A´ are offset reproductions of profiles in place, as AA, and show the apparent displacement of the sea floor along the Mendocino, Pioneer and Murray faults. (Vacquier, 1965.)

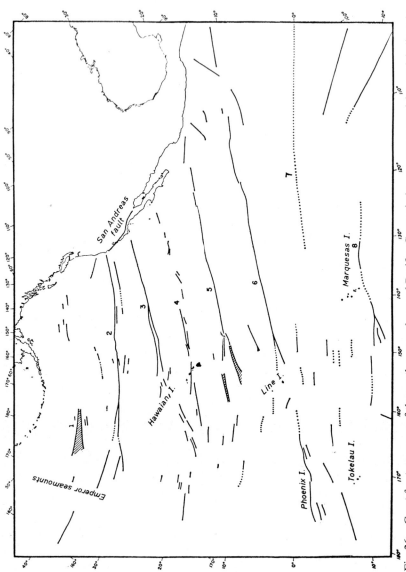

Fig. 26.   Great fracture zones of the north-east and central Pacific. Great circles are represented by straight lines in this projection (Menard, 1967a). Fracture zones: 1 Chinook; 2 Mendocino; 3 Murray; 4 Molokai; 5 Clarion; 6 Clipperton; 7 Galapagos; 8 Marquesas.

of Figure 25), ocean floor had been engulfed on the north side or produced on the south side. In this disturbed region, topographically rough and bordered on the east and west by two groups of submarine mountains just south of the fault, the north-south correlation disappears.

The magnetic alignments continue south, but are scarcely discernible beyond the next fracture, Molokai. Still farther south, the East Pacific ridge, not in evidence after Mendocino, reappears at the mouth of the Gulf of California about 25°N. More faults, perpendicular to the ridge, if not to the magnetic anomalies, have been traced by their topography. They have been named, from north to south, Molokai, Clarion, Clipperton, Galapagos, and Marquesas, the last not meeting the crest.

The great fractures of the north-eastern Pacific have been mapped by soundings out into the Central Pacific, beyond the boundaries of the East Pacific rise (Menard, 1967a); they branch out after a certain distance beyond the flanks. Figure 26, using a central projection, shows that they are approximately great circles. The most remarkable example is the Clipperton fracture, which is practically continuous as far as the Line Islands, seems to disappear there under the volcanic apron and the pelagic sediments of the group, but then reappears south of the equatorial belt of sedimentation as a lineation of trenches, rises or generally minor depth changes (however, a rise of 2000 m is found in the Phoenix Islands beside an 8000 m trench, the deepest in the Central Pacific). The total length of the fracture would amount to 9800 km.

It is hard to reconcile these observations with an earlier suggestion of Menard developing an idea of Hess. Impressed by the contrast between the Atlantic-Indian Ocean ridge midway between the continents and the non-central location of the East Pacific rise, Menard supposed that a ridge occupying the middle of the Pacific had disappeared by gradual slumping. This was the *Darwin Rise* (Darwin had studied the subsidence of coral atolls); its location roughly corresponding to the great basaltic effusions of the Pacific, which Menard compares with the plateau basalts, particularly those of Iceland. According to him, more than 100 m.y. ago, the Darwin rise towered 2 or 3 km above an ocean more than 5 km deep. Its length was 10 000 km, its width 2000 km, with longitudinal faults and local fractures giving rise to the great transverse fractures. Volcanic activity occurred much later; if this were generally true it would account for the rather moderate volcanism of the active ridges.

But it will soon be seen that the great transverse fractures are instead connected with the ancient history of the East Pacific rise.

4. Active Fractures and Fossil Fractures

The great fractures of the north-east Pacific are practically aseismic. They form in fact a highly special group; they are fossil witnesses to a former state. Their discovery was followed by that of many others, all corresponding to a lateral displacement of ridges; although they were mainly identified by their bathymetric aspect, developments in seismology make it possible at times to assign definitely to them the always numerous

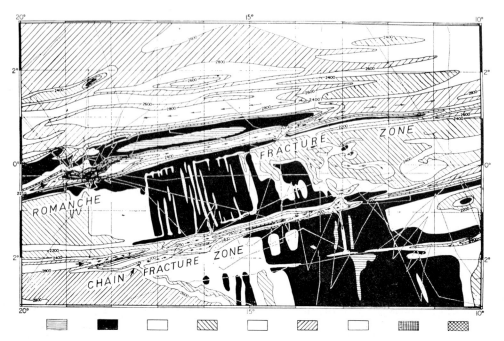

Fig. 27.   Partial depth chart of the Chain and Romanche fracture zones. Sounding lines are dotted.
Contours at 200 fathom intervals where possible. (Heezen, Bunce, Hersey, Tharp;
1964, *Deep Sea Res.* **11**, 14.)

nearby epicenters. These active fracture zones are generally much shorter than those
of the East Pacific. We may note, however, the great Eltanin fracture (shown as a
double dotted line in Figure 48) which probably stretches from the New Zealand
plateau all the way to the South Sandwich arc (Tharp and Hollister, communication
at Zurich, 1967).

Some fracture zones literally cut the ridges to pieces in some regions. The equatorial
Atlantic is scissored by a very large number of parallel fractures. As examples may be
noted the Chain fracture, which appears to move the ridge 320 km to the left, and
the Romanche, which displaces it by 860 km (Figure 27).

The Indian Ocean is similarly cut up in several of its parts, as may be seen in the
physiographic map of Heezen and Tharp, whose principal features are reproduced
in Figure 28. With two or three important exceptions, these 'transverse' faults trend
mostly north-south, as do most of the structural features of the Indian Ocean, and not
perpendicular to the general course of the ridge. Figure 29 (see p.32) has an example in
the Owen fracture cutting the Carlsberg ridge at the entrance of the Gulf of Aden. An
important fact is shown as well: the transverse fractures are aseismic (or nearly so)
except in the sections between the ridges they seem to have displaced, but these
sections then constitute the most rugged and seismically active portion of the ridges.
The map of Figure 30 (see p.33) well depicts the small active faults beside the great fossil
faults. Menard (1966) considered it possible that all ridge earthquakes might be due to

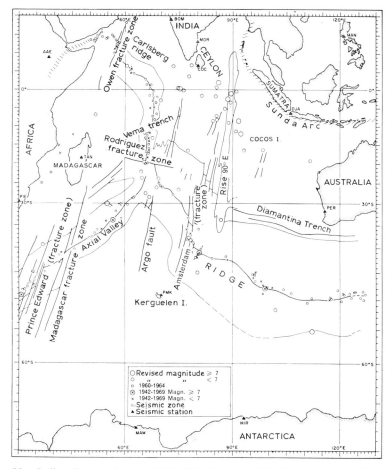

Fig. 28. Indian Ocean epicenters 1942–1964. (Stover: 1966, *J. Geophys. Res.* **71**, 2576.)

small transverse faults, and Laughton's studies on the Gulf of Aden tended to agree. The question has been elucidated by Sykes (see end of Chapter III).

In the Arctic Ocean also, as Demenitskaya and Karasik showed in 1966, fractures are found running from the Lomonosov rise to the continent which cut the Gakkel ridge transversely.

5. Return to the East Pacific Rise

The use of magnetic anomalies has led to clarification of the complex structure presented by the East Pacific rise in the region of crests and trenches north of the Mendocino fracture (40°N). The ideas of Tuzo Wilson on what he names transform faults (see Chapter III) have played a great part here, but the facts may be stated simply. Figure 31 of Talwani, Le Pichon and Heirtzler (1965) shows that the seismic belt which comes from the north-east along a fault bordered by the Queen Charlotte Islands has the form of the broken line ABCDEF. The portions AB, CD, EF, coincide

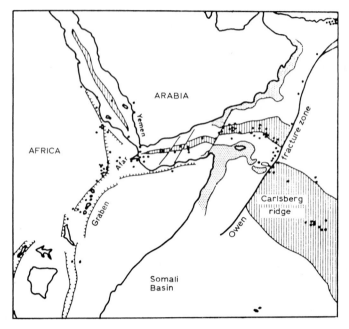

Fig. 29.   Offset of the Carlsberg ridge by the Owen fracture. (Laughton: 1967; *The World Rift System*, 80, Ottawa.)

with small known ridges called Explorer, Juan de Fuca and Gorda. A traverse of the Juan de Fuca ridge by the Vema indicates a structure similar to most ridges, but its flank seems to disappear under the continent.

On the general map of positive anomalies (Figure 13), the three small ridges, particularly Juan de Fuca, are seen to be successive axes of symmetry for the anomalies, linked by small transverse fractures.

South of Mendocino, no ridge crest can be seen. Menard considers then that one is on a flank of the ridge. The offset of the 5000 m isobath (Figure 32) supports this view. At the same time the system of magnetic anomalies is offset about 1600 km to the east (including the effect of the Pioneer fracture). The axial zone would correspond to the high part of the Basin and Range Province and the Colorado plateau, which are supposed to bear the same relation to the hidden ridge as the high African plateaus have to the graben of the great lakes.

If this point of view is accepted and the great fractures of the East Pacific are supposed to be genetically connected with the ridge, the question arises of their possible extension to the continent; this was the subject of an aeromagnetic survey by Fuller in 1964. The Mendocino fracture can be followed into Nevada as far as 115°W, but the characteristic east-west correlation between the magnetic profiles becomes weak. Much farther on, Drake *et al.* showed in 1963 the existence of a fault antedating the Mesozoic on the east coast of the United States at the latitude of Mendocino (but an observer on one side of the fault would see the other move to the right,

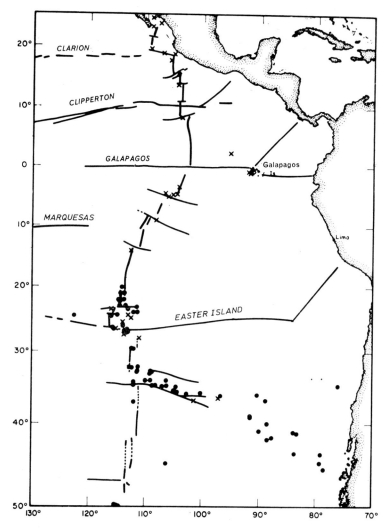

Fig. 30.   Crest of the East Pacific rise, offset by active fractures. Dots are epicenters determined by
Sykes (1963), crosses determined by the International Seismological Service. (Menard, 1966.)

whereas Mendocino shows preponderantly a left-hand slip). It extends over 600 km
of continent and then, in the sea, perhaps to the Atlantic rise. In Figure 24 it interrupts
the coastal anomalies and puts an inflexion in the magnetically quiet zone.

The Murray fracture occurs just opposite to the Transverse ranges in California
(Figure 32) with Tertiary left-hand slip, that is to say, again opposed to the apparent
displacement of the Murray.

Finally, opposite the Clarion fracture, a belt of volcanoes crosses Mexico from
west to east, and it has been supposed that Clarion and Clipperton may reach the
Antilles arc, which they may have displaced to its present position.

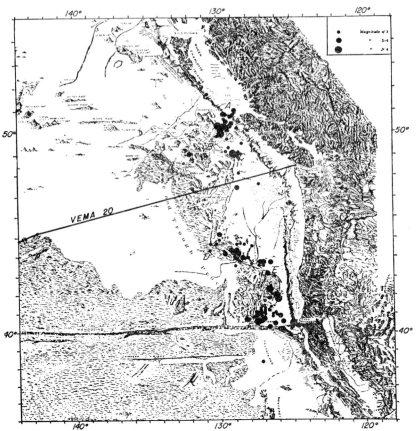

Fig. 31. Physiographic bottom chart of Menard. The epicenters of 1961–1963 are preliminary determinations by the *U.S. Coast and Geodetic Survey*. (Talwani *et al.*, 1965.)

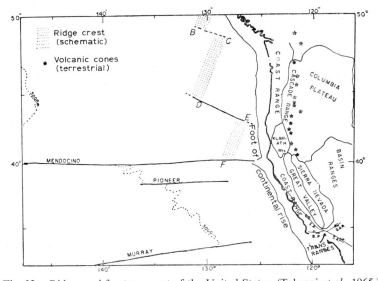

Fig. 32. Ridges and fractures west of the United States. (Talwani *et al.*, 1965.)

CHAPTER III

REVERSALS OF THE EARTH'S MAGNETIC FIELD.
VINE AND MATTHEWS' HYPOTHESIS

1. Remanent Magnetization of Lavas and the Magnetic Field in the Past

Gilbert, Queen Elizabeth's physician, discovered that the Earth's magnetic field was similar to that of a uniformly magnetized sphere; it is also equivalent to the field of a small magnet or 'dipole' placed at the centre. Actually the field varies continually; if its annual mean is taken and if it is developed in spherical harmonics as was done by Gauss, it is found to be almost exclusively of internal origin, and it is found that the leading terms, by much the largest, correspond exactly to the field of a magnetic dipole, currently inclined at 11.5° to the geographical axis.

The only satisfactory theory of the Earth's magnetic field, although by no means complete, follows Elsasser and then Bullard in attributing it to motions in the conductive liquid of the core. Whatever the precise mechanism of this self-exciting dynamo may be, and however its motion originates, (probably by thermal convection, but possibly by turbulence induced by the precession of the equinoxes) the Coriolis force due to the Earth's rotation plays a major role, with the result that the main field produced must be that of a dipole along the axis of rotation; the remainder of the field, dipolar or not, should fluctuate with time constants of the order of a hundred or a thousand years; these are characteristic of the so-called secular variation. Thus the angle of 11.5° between the magnetic and geographic axes is only temporary; the average value over a sufficiently long time, 100000 y. for example, would be very small.

These fluctuations are clearly evident in old magnetic measurements, but the latter only go back three or four centuries, and the oldest give rough values of the declination only. Fortunately they are complemented, for historic times, by the study of the magnetization of baked clays by archeologists; these studies make use of the elaborate techniques of Thellier (1966), which have served as a model for similar measurements on rocks. We cannot dwell on this science of archeomagnetism, but it does give evidence of slow variations of some tens of degrees.

The data directly useful to us comes from the magnetization taken by some rocks under the influence of the field present at their formation. This magnetization, generally parallel to the original field, is largely preserved ('remanent' magnetization) and is found today superimposed on the magnetization induced by the present field. This is what happens when lavas cool below the Curie point and acquire a generally long-lived 'thermo-remanent' magnetization, which can be isolated by delicate techniques and made to reveal the direction (or even the intensity) of the field that produced it.

Sedimentary rocks also possess remanent magnetization, usually very weak;

initially it was explained by the lining up of the grains parallel to the field during their very slow settling in the ocean. But it is probably better to consider it the effect of the magnetic field on the formation of crystals during the chemical transformation of consolidated sediments.

For all the Quaternary lavas of a given region the *direction* of the field shown by their magnetization is the same (to within 20° or 30°), which is natural if the main field is indeed that of an axial dipole and if the region has not moved rapidly with respect to the geographic pole. But the *sense* of the field may alternate. At the beginning of the century, Brunhes and David discovered an outflow of plateau basalt at Pontfarein (in the Cantal department) which they took to be Miocene but which must be Upper Pliocene; its direction of magnetization corresponded roughly with that of the present field, but had the opposite sense. A clay metamorphosed by the flow also had reversed magnetization. Shortly afterwards Mercanton, then in 1929, Matuyama added many more examples of this reversal of field. In examining 139 samples of volcanic rocks from the Quaternary and Upper Pliocene collected in Japan, Manchuria and Korea, Matuyama found a large proportion with magnetizations approximately in the sense of the present field or its opposite, with a few in random directions.

From these discoveries, it thus appeared possible that the Earth's field underwent rapid reversals over the whole globe at the same time, corresponding in the light of current views to a reversal of the axial dipole. Such an idea did not find ready acceptance. We cannot recount the successive changes of geophysical opinion, interesting as they are for the historian of science. It suffices to say that soon after Néel had found in 1951 several theoretical models by which a lava might acquire a magnetization in the reverse sense, one of these mechanisms was revealed in a dacite from Mount Haruna by Nagata and Uyeda. Other models have since been proposed, notably those of Verhoogen in 1962, which were particularly embarrassing to the proponents of simultaneous reversals on a global scale because they corresponded to a very slow change of magnetization.

Nevertheless the determinations carried out by teams of paleomagneticians on tens of thousands of samples dating back tens of millions of years have found many instances of concordant double magnetization as at Pontfarein, and more important still, have found many agreements in the date of reversals observed in different places in rocks of differing composition.

Today it seems clear that self-reversal involves a very rare category of rocks and minerals. Although the evolution of the field during a world-wide reversal is still obscure, the existence of these reversals can no longer be denied, and their frequency of occurrence is also fairly well known, at least for the last few million years. For example, Dagley *et al.*, (1967), in studying a thousand basalt flows in the east of Iceland which were hard to date but corresponded to a total thickness of 9 km and a duration of about 20 m.y., found about sixty reversals among them, representing 5% of the time. Statistically speaking, the field intensity during transitions would amount to something like a third or a quarter of the mean intensity during the normal or reversed states (much larger temporary fluctuations would occur).

2. Periods of Polarity and Intervening Events

While the earliest work on reversals, such as Hospers' study on the Icelandic flows, and Roche's on those of Auvergne, sought to establish relationships with stratigraphy, such comparisons were soon found inadequate because of the frequency of reversals; from 1963 an increasing number of groups have practised radioactive dating by the potassium-argon method.

A synthesis which was to have important consequences for the problem of sea-floor spreading was undertaken by Cox *et al.* in a series of papers in 1963. These authors began by recognizing three major periods of polarity, that is, with a definite

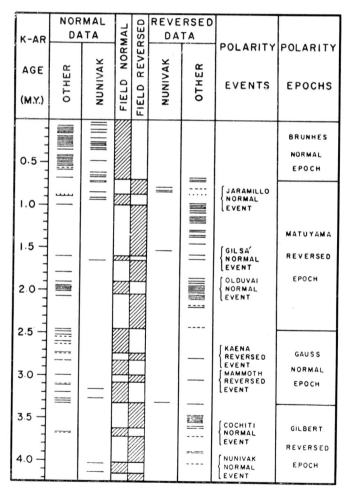

Fig. 33. Magnetic polarity and isotopic ages by the potassium-argon method; results from 29 basalts of Nunivak I. in the Behring Sea, together with earlier ones. Each line represents an age and polarity determination (*dashed*, for rocks containing glass; *dotted*, for magnetization in intermediate directions) (Cox and Dalrymple, 1967.)

sense for the axial dipole field, all having durations of the order of a million years. The current period, which they later named the *Brunhes* epoch, is characterized *in general* by normal polarity, identical with the present sense. During the preceding epoch named after *Matuyama*, polarity was generally reversed; before that it was mostly normal during the *Gauss* epoch, and finally reversed again in the *Gilbert* epoch, whose commencement is still ill-defined.

Then in 1964, the results found by various laboratories among the volcanic rocks of North America, Hawaii, Europe and Africa made it necessary to intersperse brief episodes into the body of the foregoing *epochs*, named after geomagneticians, during which the polarity had the opposite sense. Cox *et al.* (1964) called them *events* and named them after the place of origin of the corresponding samples. Two events were discovered in 1964; the normal *Olduvai* event (after the well-known paleontological site) occurring in the Matuyama epoch, and the reversed *Mammoth* event during the Gauss epoch.

This distinction between long epochs and short (order 10000 y.) events arose from theoretical considerations soon to be discussed. It is beginning to appear rather awkward, for the number of recognized events continues to grow. At present the distribution of the intervals of polarity is becoming a regular one, with the probability falling off with the duration (Figure 51).

Newer events, all with place names, are *Jaramillo*, introduced by Doell and Dalrymple in 1966; *Gilsa*, rather doubtful, by McDougall and Wensink; *Kaena*, by McDougall and Chamalaun, also in 1966; *Cochiti* and *Nunivak*, by Cox and Dalrymple (1967). Figure 33 shows the state of knowledge at the time of their article. They think that modifications to the chart of reversals will happily become fewer, but this seems optimistic. So far the Brunhes epoch has not been blemished by any events; if exception is made of one still uncertain observation (Bonhommet and Babkine, 1967).

3. Magnetization of Ocean Sediments and Various Phenomena Connected with Reversals

Far from the coasts, oceanic sediments are laid down at an average rate on the order of a meter per m.y. Cores 20 m long can be secured in the great deeps, although not readily, with piston corers. In them are continuous records of events of the Quaternary and the Lower Tertiary, but for want of isotopic dating, the tempo of these phenomena is only approximately known, as it depends on the sedimentation rate. We shall see that assuming parallelism between the remanent magnetization and the fossilized field leads to results agreeing with those from lavas.

It appears that Harrison and Funnell in 1964 were the first to recover certain reversals on this assumption. In an extensive study, Opdyke *et al.*, (1966), using Antarctic cores, were able to disregard directions in azimuth because of the steep dip of the magnetic field; with the help of a paleontological stratigraphy based on radiolaria they identified the epochs of Brunhes, Matuyama and Gauss, and also the Jaramillo,

Olduvai and Mammoth events. Other events were identified later. Not all cores show the complete sequence of reversals; it even appears that in other oceans rather few cores have clearly-defined reversals. Nevertheless the method is a useful complement to lava studies.

In another application, Opdyke and Wilson sought to verify the mean axial dipole hypothesis. Among 270 cores from all the oceans they chose the 56 in which the dispersion about the mean inclination was less than 15°, and then compared this inclination with that expected from the axial dipole. Agreement was excellent for the Brunhes and Matuyama epochs. The mean pole found on the assumption that the magnetizations were in the meridian plane was only 1° from the pole of rotation in the Brunhes case.

The faunal zones marked φ, χ, ψ, Ω in Figure 34 correspond to different populations of radiolaria, and seem to be related with the periods of reversal (Figures 34 and 35). The transition χ–ψ coincides with the disappearance of numerous species and the appearance of new ones (Figure 35), and this chimes with Uffen's idea of 1963 that if the magnetic field is weak during a reversal, the unimpeded arrival of cosmic radiation could cause mutations capable of modifying the course of evolution. The numerous discussions (Waddington in 1967, Black in 1967, Harrison in 1968, etc.)

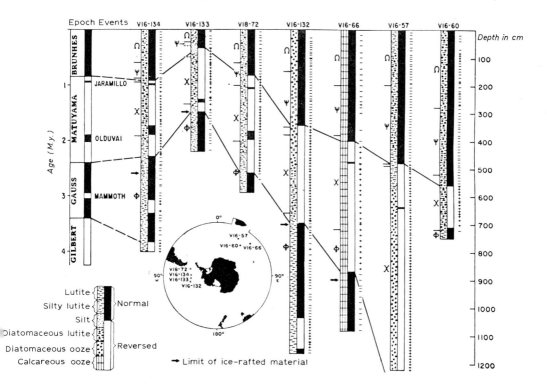

Fig. 34.   Magnetic correlations among seven Antarctic cores. Ages in m.y., depths in cm. Minus signs indicate normally magnetized specimens; plus, reversed. Greek letters denote faunal zones. (Opdyke *et al.*, 1966.)

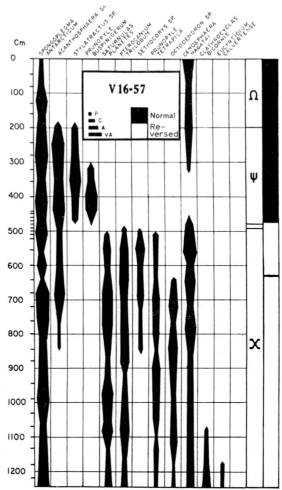

Fig. 35.   Abundance of indicator species in an Antarctic core. Scale at left shows depth within core. (P = present; C = common; A = abundant; VA = very abundant.) (Opdyke *et al.*, 1966.)

that this idea gave rise to among physicists and biologists will not be recounted. But it seems that exposure of the atmosphere and oceans to cosmic or solar particles with the characteristics of the present radiation would alter the mutation rate by only a few percent. An improbable indirect influence has been suggested, through the effect on climate.

One last coincidence between field reversals and a phenomenon apparently totally unrelated to them has been noted. In several parts of the Earth are found widely strewn glassy meteoritic particles called tektites. Potassium-argon dating and the fission-track method indicate a common date of origin of −700000 y. for certain swarms (the tektites of Australia, Indonesia, Indochina and the Philippines) even though their morphological differences are marked. But that is the date of the change from Matuyama to Brunhes.

There are two hypotheses on the origin of tektites; they may come from the Moon or be the result of the impact of large meteorites on the Earth. Glass and Heezen (1967) opt for the second hypothesis, and suppose that the Far Eastern tektites (whose mass might total 150 million tons if the microtektites discovered by Glass (1967) in cores taken off Australia are included) are the debris of a great meteorite which exploded near the ground after having broken into several fragments. The fall of this meteorite is supposed to have triggered the reversal of field. Other such impacts, whether or not they produced tektites, gave rise to other reversals. But the Tunguska meteorite of 1908, probably caused by the explosion of the nucleus of a comet, although it showered an area of 2000 km^2 with tiny spheres of siliceous glass or of metal, is supposed to have provided insufficient energy for a reversal. Such ideas imply that the geomagnetic field can reverse itself with disconcerting ease. Without seeking to draw a parallel between the earth and certain stars whose polarity changes in a few days, let us see if there are any reasons for believing in such instability.

4. Theoretical Considerations on the Origin of Reversals

The problems in hydromagnetic theory posed by the Earth as a self-excited dynamo are extremely complex, and their solution so far is barely outlined.

Simple models, consisting of solid conductors in principle, are not physically realizable with one exception, but have had some success because they can be made to produce numerical results. We mention only the disk dynamo suggested by Bullard in 1955, and the set of two coupled identical disk dynamos studied by Rikitake in 1958 (Rikitake, 1966; or Rikitake in Runcorn (ed.), 1967). This latter, schematized in Figure 36, has been the subject of numerous calculations. The two dynamos are

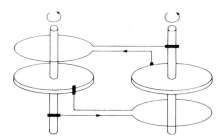

Fig. 36.   Pair of coupled disk dynamos. (Rikitake, 1958.)

supposed to be driven by constant equal couples. In these conditions it may be shown that the two rotation speeds differ by a constant, but the speeds themselves are highly variable and the output currents still more so. Figure 37 based on Allen's calculations shows two characteristic examples. In Figure 37a, the current in the first dynamo, and thus the corresponding field as well, reverses itself during very short intervals, analogous to 'events'. In Figure 37b, the same current fluctuates about values alternately positive and negative, analogous to terrestrial 'epochs'. In either case, self-

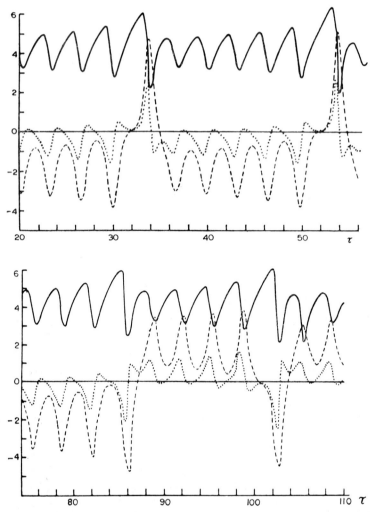

Fig. 37a and b. Oscillation of a pair of coupled disk dynamos for particular values of the defining parameters. *Full line*, variation of the rotation; *dashed*, current in first dynamo; *dotted*, current in second. (Allan, 1962.)

excitation does not lead to a stable state with energy losses equalling the output of the driving couples, but to an unstable state with increasing amplitude of oscillation leading to reversal.

Just before the reversal, when the field is very weak, small perturbations can alter the action of the dynamo so as to maintain or reverse polarity in a sense contrary to the theoretical predictions. Collision with a large meteorite may be considered, as we have stated, as one of the perturbations triggering a reversal. This is not the likeliest cause; in the terrestrial dynamo, the motions are more or less turbulent. Perturbations due to turbulence are likely to be important because they act on the magnetic field within the core, where it is much greater than one might suspect from the field at the surface. (In fact, the growth of the field perceived on the surface, known as the poloidal field because its lines of force lie in the plane of the meridian like those of a dipole, involves the simultaneous existence of a much stronger field whose lines of force are circles centered on the polar axis, a toroidal field.) In this fashion the aleatory nature of periods of polarity could be explained.

5. Vine and Matthews' Hypothesis

Two years after the impact of the ideas of Hess on the renewal of ocean floors (presented in 1961, immediately followed by Dietz, although their publications were in the opposite order), development of these studies was greatly stimulated by an insight of Vine and Matthews (1963). Bands of magnetic anomalies parallel to the ridges had been observed in the North Atlantic, in the Antarctic, and in the Indian Ocean, particularly in the Carlsberg ridge by an English expedition. They had been attributed to the contrasting magnetization of superficial rock or even to faulted structures. Vine and Matthews related these observations to the alternations of magnetic polarity, the chronology of which Cox *et al.* were beginning to establish, and to the idea that the oceanic crust issues from the ridges and is drawn by convection towards the neighboring continents. They assumed that the lavas exuded in the axis of the ridges took on the magnetization of the then current field on cooling, then moved aside to make way for new flows, which became magnetized in the same sense, and so on, until a reversal of the sense of the inducing field caused the new lavas to become magnetized in the opposite sense, thus producing a separation of the older axial zone into two symmetric blocks. Representing the masses which produce the first few anomalies schematically by vertical blocks of alternating magnetization, Vine and Matthews succeeded in reproducing the main features of these anomalies.

The lava layers responsible for the magnetization, certainly very irregular, are not necessarily very thick. Samples of fresh basalt dredged up from the ridges, usually pillow lavas, display a very high remanent magnetization, of the order of several hundredths of an electromagnetic unit per cubic centimeter, with a very low Curie point, around $140°$ to $180°$, according to Petrova and Pechersky (communication at Zurich, 1967), which corresponds to a magnetizable depth of less than 2 km if the thermal gradient on the ridges is taken to be about three times the usual gradient of $30°/km$ (see Chapter VII). The induced magnetization of these same basaltic lavas when cooled is slight, say of the order of 2% of the remanent magnetization, compared with around 10% for usual volcanic rocks and 100% for well-crystallized igneous rocks (Vogt and Ostenso, 1966; Luyendyk and Melson, 1967, Opdyke and Hekinian, 1967). All these properties appear to be due to the relatively large proportion of titano-magnetites in the basalts in question together with the particular conditions of their cooling.

It was seen in Chapter II that the central anomaly was much more pronounced than the lateral ones (around twice as strong). To explain this, Vine and Matthews suppose that the magnetization of the two lateral blocks has been contaminated, after the reversal which separated them, by subsequent non-axial volcanic flows. An example of this phenomenon was provided by the preliminary drilling for the great American Mohole project (now abandoned), which encountered reversed basalts in a positive anomaly region.

The question has since been taken up again by Matthews. Following the Icelandic model of Bodvarsson and Walker (see Chapter II) he supposes that the magnetizable materials are injected into the axial regions of the ridges in the form of very thin vertical dikes which displace the dikes previously emitted, and that the probability of the injection taking place at a given distance from the axis falls off rapidly with the distance. By using the simplest model, that of identical dikes whose distance from the axis follows a normal Gaussian distribution, it is found that the great majority of injections must occur very close to the axis; in fact the standard deviation reproducing the profiles observed in the North Atlantic around 45° is under 5 km, the half-width of the axial valley (Matthews and Bath, 1967).

Eruptions from fissures or minor lava flows may be invoked, as well as dikes. But larger outflows or volcanic structures upset the parallelism and symmetry of the magnetic anomalies rather than contributing to them.

While contamination of the lateral anomalies surely plays a part in the predominance of the axial anomaly, its effect seems insufficient (Pitman et al., 1968). Perhaps the effect of induced magnetization is important (Thellier). Temperatures are certainly high in the axial zone, and it appears that the susceptibility of a magnetizable substance is very great a little below its Curie point. This is so, at any rate (Ghorbanian in 1966) for a magnetite as long as it is not more than about 100° below its Curie point of 580 °C.

As elsewhere in this treatment the chronological development of the hypothesis of Vine and Matthews will not be followed in detail. It may be mentioned, however, that in 1965 a first close comparison of magnetic profiles and the time scale of reversals allowed Vine and Wilson to arrive at a spreading rate of 1.5 cm/y. for the Juan de Fuca ridge; this was correct only in order of magnitude, but results have since greatly increased in number.

6. Transform Faults

The theory of sea-floor spreading has led to a better understanding of the respective roles of the ridges and their transverse fractures. Consider (Figure 38) two successive segments AA', BB' of the crest of a ridge on either side of a fracture perpendicular to them, as is often the case. Each segment is the seat of an expansion which gives rise to relief and symmetrical magnetic anomalies on both sides. We shall see that the spreading rate changes little over great lengths of ridge; it is sensibly the same for two consecutive segments. The topography and anomalies on either side of the fracture can thus be brought into superposition by moving the two segments into alignment, as if they had slid apart, but the relative motion of the two sides of the fracture, in the portion A'B between the crests, is in the opposite sense to that of the apparent strike-slip movement.

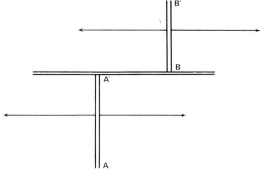

Fig. 38.

It is on this segment A'B where the expansions oppose each other and where the relative motion of the two edges is large, that earthquakes might be expected to occur, and indeed this is what is observed.

Such an interpretation of transverse fractures is perfectly suitable for those of modest length which intersect the ridges in their curved portions. One might hesitate to apply it to the fossil fractures of the Pacific where it would imply an expansion over thousands of kilometers from a now vanished ridge. On the other hand it may be recalled that the extensions of great fracture zones under the American continent show reversed displacements, as would be natural if one were on the other side of the ridge.

Transverse fractures, as we have just interpreted them, are the simplest example of what Wilson (1965) has named *transform faults*. According to him, none of the great geological structures – ridges, island arcs, or on land, grabens, arcuate ranges, terminates in a free end. The deformation corresponding to it (an extension for the ridges and generally a compression for the arcs) must be transmitted to the following structure through a fault which, for example transforms an arc into a ridge (one might better say that it makes the transition from arc to ridge). It has just been seen how transverse fractures are faults transforming one dorsal into another. The line of epicenters running from Bouvet Island on the mid-Atlantic ridge to the far end of the South Sandwich arc (Figure 2) is an example of transformation from ridge to arc.

This concept of transform fault or fault of transition has proved very fruitful even though the hypothesis as to the continuity of major structures is at times found wanting.

7. Focal Mechanism of Ridge Earthquakes

The best confirmation of Wilson's views was found through study of the direction of first motion in those earthquakes, all shallow, whose foci lie on the ridges. This work was carried out by Sykes (1967) with the accuracy achievable with the world-wide American network of identical long-period seismographs. Fewer than 1% of the data departed from a quadrantal distribution, in contrast to 15% in earlier investigations.

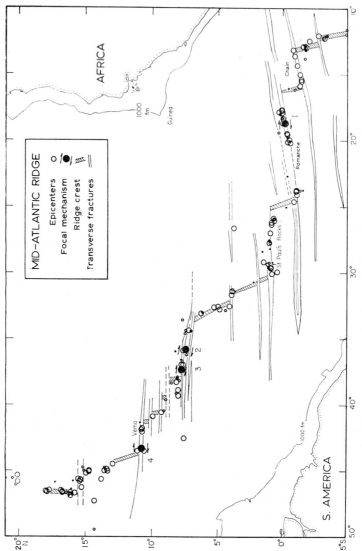

Fig. 39. Equatorial Atlantic epicenters 1955–1965, with direction of first motion and fault plane trace shown for four of them. The best determined epicenters are shown by circles, others as dots. (Sykes, 1967.)

An example of Sykes' results for the equatorial Atlantic is shown in Figure 39. Almost all the epicenters on transverse fractures are found between the segments of the crest. In four cases the direction of first movement has been determined, and if it is admitted, as seems logical, that the fault plane is the plane of the fracture rather than the vertical plane perpendicular to it, this direction corresponds to the model of Figure 38, and not to the hypothesis of strike-slip motion separating the segments. Sykes obtained similar results for a small offset of the Atlantic ridge just north of Iceland, and for three offsets of the East Pacific rise, particularly for the small Rivera fracture, just south of the Gulf of California, which extends the Clarion fracture, but at an angle of 20° to 30° to it; a possible explanation will be seen later.

Nevertheless in Figure 40, two earthquakes (numbers 6 and 7), in a well-studied region where any offset of the ridge larger than a few tens of kilometers would have been evident, give first motions towards the epicenter for all distant stations, corresponding to a normal fault parallel to the axis. To within the precision of determination the maximum stress corresponds to an extension. One earthquake of the Arctic ridge and three of the African grabens give similar results; however Sykes has found none such for the East Pacific rise.

The definiteness of Sykes' results, in spite of the restricted number of cases examined, suggests that all ridge earthquakes fall into two classes, those of transverse transform faults, and of normal faults of the axial graben.

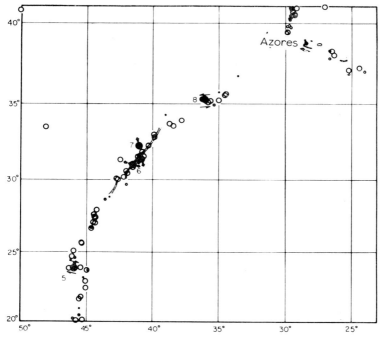

Fig. 40.   Epicenters of 1955–1965 on part of the mid-Atlantic ridge, with focal mechanism for four of them. Median valley hatched. The thick arrows indicate the directions of greatest tension for earthquakes 6 and 7. (Sykes, 1967.)

CHAPTER IV

OCEAN SPREADING

1. From Ridges to Trenches

If the hypothesis of sea floor spreading from the ridges is admitted, what happens to the floor thus produced? An initial reply may be given in very general terms. Firstly by pushing aside the continents, the spreading may enlarge the oceans containing the ridges; some geophysicists consider this the rule and ocean spreading to be merely a reflection of a general expansion of the Earth. But there are serious reasons for considering the radius of the Earth to be nearly invariable, and so the existence of regions where ocean floor disappears must be admitted. These regions would correspond to trenches and continental edges, as Hess and Dietz thought as early as 1961. It has been seen in Figures 23 and 24 that linear anomalies are there observed which would represent the final stage of anomalies parallel to the ridges.

The paper of Oliver and Isacks (1967) on deep focus earthquakes of the Tonga arc supports the by now old views of Hess and Dietz. The same earthquake, enregistered at the Tonga Islands, on the arc, and at Fiji, inside the arc (Figure 41), showed great differences in amplitude and period. Their explanation is as follows: the displacement of a thin magnetized basaltic sheet without deformation is only conceivable if the sheet is incorporated in a relatively thick resistant layer, of the order of 100 km, say. Oliver and Isacks adopt the classic term *lithosphere* for this layer in which seismic waves undergo very little attenuation. On the other hand, they suppose the lower layer or *asthenosphere* to be capable of slow flow; over the short periods characteristic of earthquakes it would act as an imperfectly elastic solid with wave amplitude falling off with distance. The asthenosphere is supposed to drag along the lithosphere, forcing it and the crust to plunge under the island arc

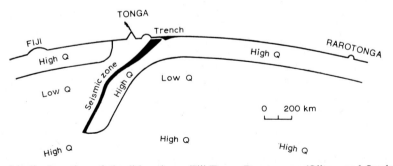

Fig. 41. Idealized section of the lithosphere, Fiji-Tonga-Rarotonga. (Oliver and Isacks, 1967.)

with the production of earthquakes, at first of shallow, then of deep focus (Figure 41) in a zone less than 25 km thick.

One measure of the ability of a medium to transmit waves is its quality factor Q, such that the relative loss of energy per cycle is $2\pi/Q$. What has been said earlier amounts to saying that Q is high in the lithosphere, low in the asthenosphere. The path of the waves from a deep earthquake to the Tongas lies entirely in the first medium, the path to the Fijis largely in the second, which explains the appearance of the respective seismograms. Katsumata found similar results for the Japanese arc up to 350 km.

The model of Oliver and Isacks has been most successful. Adopting it, two modes of origin of the ridges are to be distinguished, according to Hess. In the first case, they begin as a fracture in the middle of a continent and grow as the edges separate, thus creating a new ocean; the ridge is mid-oceanic, as the mid-Atlantic one. This idea fits in readily with the fact that the Atlantic chains are transverse with respect to the coasts. The Indian Ocean example is much more complex; continental fragments like the Seychelles appear to have subsisted in the midst of the newly formed ocean. In the second case the ridge is born in an ocean already constituted, stabilized, or even in regression; sea floor spreading then comes to an end in the trenches bordering the continents or in island arcs of the Tonga type. The East Pacific rise in its westward development is the best present example of the second mechanism.

All this provides a valuable clue among the facts whose complexity we shall discover progressively.

2. Simple Examples of Sea Floor Spreading

The first chapter has provided an overall view of seismic geography. But transform faults and ridge segments give a fine structure to the shallow focus earthquake lines; moreover there are considerable regional variations in the magnetic anomalies and ridge morphology. It would be appropriate now to be able to analyze local details. All the oceans are being reviewed in this way, and there are already many characteristic examples.

A notable article by Vine (1966) musters the cases then explicable and shows their agreement with the chronology of reversals (which at that time only included as events Jaramillo, Olduvai and Mammoth). Two of Vine's examples already considered here (Figures 16 to 19) had been studied in detail by Pitman and Heirtzler (1966): the South Pacific and the Reykjanes ridge (Figures 42 and 43). The symmetry of the profiles will be remarked, as well as the precision with which they are represented by the magnetization of blocks of alternating polarity. The blocks are the elements of a layer 2 km thick upon the sea floor magnetized with an intensity of ± 0.005 G in the direction of the present field. Beyond the 4 m.y. limit of the reversal scale the Pacific profile is well defined enough to suggest further reversals; *if the spreading rate were supposed constant*, these would in turn be dated. Although this hypothesis is not at all obvious it has allowed Pitman and Heirtzler to apply with some success

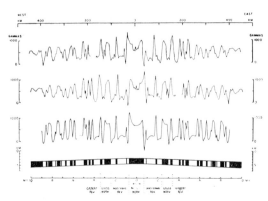

Fig. 42.   Middle curve is the profile of magnetic anomaly given by *Eltanin*-19, with above it the same profile turned end for end (west on right), below it the magnetic profile calculated from the model of the Pacific-Antarctic ridge shown at the bottom. The time scale in m.y. corresponds to a spreading rate of 4.5 cm/y. Shaded regions normal, remainder reversed magnetization.
(Pitman and Heirtzler, 1966.)

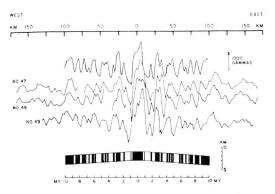

Fig. 43.   At the bottom, the Pacific-Antarctic model applied to the Reykjanes ridge with a time-scale corresponding to a 1 cm/y. spreading rate. Top profile is calculated from the model. The three intervening profiles are from the airborne survey, and are projected perpendicular to the ridge axis.
(Pitman and Heirtzler, 1966.)

(Figure 43) the scale so prolonged to the Reykjanes ridge, in spite of the differences in rate (1 cm/y. as against 4.5 cm/y.).

Vine (1966) also examines the Red Sea case, shown schematically in Figure 22. It is already less simple. Floor spreading at 1 cm/y. would account for the magnetic anomalies on two profiles, but the edges of the axial valley would have different ages on the north and south profiles, so that normal floor spreading from a tear in the crust is not satisfactorily admissible. Probably the initial phenomena were complicated. On the other hand there may be some doubt about the outline of the anomalies

(Figure 21). The Red Sea and the Gulf of Aden (Figure 74), as well as the Gulf of California (Figure 59), probably correspond to an oblique floor spreading stemming from en echelon ridge segments linked by transform faults.

These suffice as examples of immediate success for the scheme of Vine and Matthews. This generally occurs with the anomalies of a crest provided that their extension is closely parallel to it, that is, provided that they are not perturbed by fractures or volcanic structures.

3. Expansions Without Ridges

The difficulty with the following examples is not a matter of a want of definition in the anomalies; the ridge itself is ill-defined.

In the equatorial Pacific north of the Galapagos, Herron and Heirtzler (1967) have found east–west anomalies with a certain if imperfect symmetry about the principal one; it exceeds 1000 γ as against only 200 γ on the axial anomaly of the East Pacific rise in the same latitude; this is explicable by their differing orientations. The topography is rough, sediments are absent, but nothing in the bathymetry reveals the presence of a ridge except perhaps for a very slight swelling over the magnetic axis. Seismicity in the region seems rather to be connected with fracture zones. A spreading rate of 3 cm/y. gives a fairly good representation of the anomalies.

These do not extend as far as the crest of the East Pacific rise, which has a bend in the same latitude. Herron and Heirtzler suppose that the divergent expansions on either side of the bend have created a zone of secondary extension and expansion responsible for the anomalies, as in the diagram of Figure 44. The same interpretation was subsequently used by Le Pichon for the Tuamoto rise and the Chile ridge; confirmation would be desirable.

Fig. 44.

An important case is that of the Labrador or Baffin Sea. Figure 45 taken from Godby *et al.* (1966) shows two zones of parallel magnetic anomalies lying on either side of a seismic line whose activity, according to Sykes, quoted by Johnson and Pew (1968), is about one-tenth that of the usual ridge, and which coincides more or less with a submarine canyon. The distances between anomalies, of the order of 20 km, are compatible with a reasonable spreading rate. Vine (1966) assumes that an ancient ridge which is now no longer spreading but still retains considerable seismicity, has been buried there. The complete absence of an axial anomaly, which ought to be large

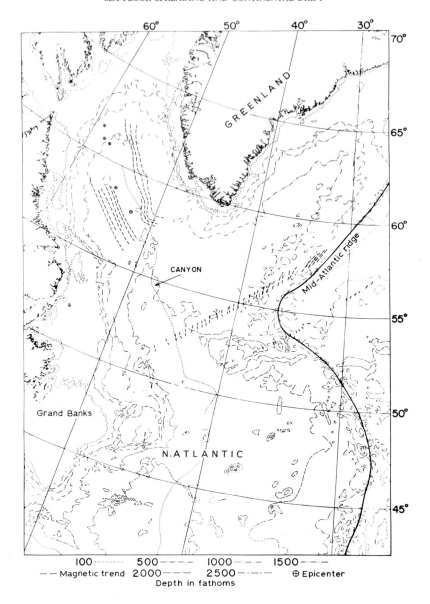

Fig. 45.   Bathymetric map (in fathoms) showing linear magnetic trends in the North Atlantic.
(Godby *et al.*, 1966.)

at this latitude, could possibly be explained if the end of the expansion happen to
coincide with a field reversal. But there are other troublesome details – acoustic
sounding shows a crest underneath the sediments of the eastward magnetic zone.
Johnson and Pew do not accept this as one of the flank ranges of the ridge, because
the other range would have to be deeply buried under the precontinental rise, and

the axial valley would be abnormally wide; they claim it to be the ridge itself, which seems even less acceptable.

Figure 45 also shows a most instructive profile carried out farther south in 1963 by the U.S. Naval Oceanographic Office using three ships travelling together at 10 mile spacing. Anomalies are seen which have much the same direction as the foregoing ones as far as the near vicinity of the Atlantic ridge, where they change direction sharply to parallel it. The influence of the enormous Atlantic ridge accordingly only extends a very short distance here.

The Indian Ocean ridge is perhaps an analogous case. The northern branch (Carlsberg ridge) was the second ridge to be identified; the southeastern branch is indisputably active; the Ninetyeast rise, in the form of a horst, seems to bear the same relation to it as the Walvis and Rio Grande rises do to the South Atlantic ridge (Le Pichon and Heirtzler, 1968). Schlich and Patriat (communication at Saint-Gall, 1967) have observed a remarkable correlation among two 500 km long profiles cutting this branch between 140°E and 145°E, another profile around 75°E and lastly a profile of the Eltanin cruise at 120°W, in all over an arc of about 10000 km girdling the Antarctic Ocean.

If this south-eastern branch is connected with the Carlsberg ridge, although there seems to be a gap between the Rodriguez fault and the Vema zone, then the group of all presently active ridges would comprise two distinct lines – from the Red Sea to Alaska, and from the mouth of the Lena to the middle of the South Atlantic. This will be further discussed in Chapter VIII in connection with 'tennis-ball convection'.

By contrast, the ridge's south-western branch, fractured to an extent seen in Figure 28, does not always show an axial valley or anomaly. Vine (1966) again considers

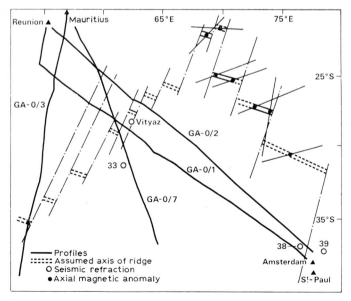

Fig. 46.   Southwest and southeast branches of the Indian Ocean ridge. (Schlich and Patriat, 1968.)

it to be an inactive ridge. However seismicity is very high there in places. It has been thought the seismicity could be related to the fractures; it would then be necessary to deny them the role of transform faults. Unlike the Labrador ridge, no pattern of linear anomalies has been identifiable in this southwestern branch. Le Pichon and Heirtzler (1968) however note that a slow rate of spreading gives a high relief, perturbing the anomalies (cf. Chapter VI); these would become unrecognizable moreover if the rate fell below 0.3 cm/y., with thickness of the intrusions under 3 km.

The Maldives and Laccadives arc, bounded on its exterior by the Chagos trench, appears to be an extension of the south-west branch, but the relation between the two structures is by no means obvious.

On the contrary Schlich and Patriat (1968), in Figure 46 give a hypothetical outline of the southwestern branch, treated as being active, composed of segments parallel to those of the southeastern branch, of which it would be a continuation, the whole not crossing the Rodriguez fault to the north.

As well as the two preceding examples of ridges where emission may have ceased, the possible existence may be mentioned of a similar ridge in the South Sandwich arc (Griffiths *et al.*, communication at Zurich, 1967), and several others in the Arctic Ocean (Johnson and Vogt, 1968).

4. Local Changes in the Rate of Expansion

If ridges have become inactive, fluctuations in the spreading of the active ridges are to be expected. The situation found by Phillips (1967) on the mid-Atlantic ridge around 27°N will serve as an initial example. Fractures are few and the crestal zone shows very good correlations among ten or so transverse profiles. Using the time-table of reversals of Pitman and Heirtzler and adopting a rate of 1.25 cm/y., anomalies are well accounted for out to 75 km or 6 m.y. Farther out, on the flanks, the anomalies remain large, their amplitude reaching as much as 30 to 50% of the axial value at around 150 km. But they are more widely spaced than those of the model. If a spreading rate of 1.65 cm/y. is assumed for this region, much better agreement is found, and the scale of Pitman and Heirtzler can be followed in this way to its limit (10 m.y.), thus dating the distant anomalies. An analogous change (from 1.4 cm/y. to 1.7 cm/y.) appears less clearly on the same ridge around 22°N (Van Andel and Bowin, 1968).

The question thus arises whether the scale itself is not vitiated by variations in the rate of spreading. This most important problem is not yet settled. The correspondence between distant anomalies in various oceans is the object of a major American program of coring and dating sea floors, the JOIDES program, but the accuracy of radio-active dating (order of 10%) is inadequate for distinguishing reversals at short intervals. And further difficulties will be encountered.

An important instance, again examined by Vine (1966), concerns the East Pacific rise, to which we must recur. To begin with, the floor spreading centered on the Juan de Fuca ridge (Figure 13) does not agree with the reversal scale unless allowance is made for the displacement of the anomalies by the faults, and also for slight changes in their direction. The movements of the blocks situated between the faults appear rather complex. The rate of expansion, this time compared with that of the South Pacific considered by Vine to be uniform, seems to have decreased from 4 or 5 cm/y. to 2.9 cm/y. 5.5 m.y. ago.

Vine suggests that as a first approximation the recent geological history of the region corresponds to Atlantic sea floor spreading driving North America over a now disappeared trench system (whose existence is considered to be a consequence of that of the East Pacific rise), and then over the crest itself.

According to Vine, the expansion centered on the East Pacific rise took place in the east–west direction giving rise to the north–south anomalies. The great fractures were transform faults intercepting the ridge and continuing beyond the now-buried crest, this explaining the change in sense observed on their continental extensions. In the course of the last 10 m.y. this expansion was superseded by a south–east to north–west movement parallel to the San Andreas fault. Trace of it is found in the anomalies themselves and also in the angle between Clarion and Rivera mentioned in Chapter III. North of Mendocino the change in direction has progressively suppressed the primitive ridge, breaking the block bearing the anomalies, and forming the new ridge in three segments. South of Mendocino, where the displacement currently follows the San Andreas fault itself, the location of the ancient crest may be recovered by starting from the crest position north of Mendocino before burial and then making allowance for the displacements measured farther to the south by Vacquier on the Mendocino and Pioneer faults. In this way Vine concludes that the ancient crest ran between Utah and Arizona, under the Colorado plateau.

South of Murray, Vine recognizes no recent modifications of the crest.

To sum up these views of Vine, the anomalies of the North East Pacific are related to the ancient crest, the great Pacific fractures are fossil, and the Utah belt of epicenters is a mere seismic relic like that of some old foldings. In assuming with Vine that the rate of spreading has always been 4.5 cm/y. north of Mendocino for the Gorda ridge, an age of 80 m.y. is arrived at for the most distant anomalies (168°W) found by Raff in 1966. The East Pacific rise accordingly would have begun its activity at the end of the Cretaceous, probably at the time when the Darwin rise was sinking, if it ever existed.

5. Worldwide Variations of Expansion

Upon the flanks of a ridge the anomalies change, often rather abruptly, to larger amplitudes and wavelengths. This change occurs at distances from the axis roughly proportional to the expansion; it corresponds to about 25 m.y. according to Vine (1966), who interprets it as an increase in frequency of reversals from this date on. Certain magnetized strips would have become too narrow to appear singly but would reduce the apparent intensity of the nearby strips. The amplitude of the inducing magnetic field may also have decreased simultaneously.

An increase in the expansion rate on all the ridges taken to be interconnected would come to the same thing as this increase in frequency, but would seem less likely (one extreme case of change in frequency is known: paleomagneticians have not found any reversals throughout the Permian and part of the Upper Carboniferous). There is no need for lengthy argument, for the simultaneity of the phenomena and the date of 25 m.y. seem uncertain.

Ewing and Ewing (1967) have brought forward arguments in favor of a truly world-wide variation in the rate. It has been seen (Figure 10) that the sediments become thicker on the flanks of the ridges. To be able to connect these differing thicknesses with differences in the speed of deposition it is indispensable to be sure that all the sediments involved are of the same nature. As a matter of fact only *pelagic sediments* are found on the ridges, in part formed in place, in part transported by winds or surface currents. Up to a basement depth of about 4 km they are calcareous muds of essentially organic origin; beyond that, the speed of deposition is less than the speed of solution of the limestone, and the only remainder is the much less thick terrigenous part constituting the red clay of the great deeps. But the region where the Ewings observe the change in thickness is still in the calcareous zone. This change, less ancient than Vine's, is thought to coincide with the transition from crest to flanks, and with a change in aspect of the system of magnetic anomalies.

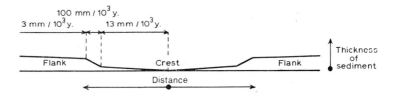

Fig. 47.   General sketch of the thickness of sediments on ridge crest and flanks, from equatorial Pacific data. Slopes correspond to the accumulation rates given if the sea-floor spreading rate is a constant 4.5 cm/y. (Ewing and Ewing, 1967.)

Ewing and Ewing make the following suggestions, schematized in Figure 47. During an initial period, which perhaps corresponded to the emplacement of the continents, sea floor spreading, whether or not it was continuous, was sufficiently general to sweep the Pacific clean of all the Palaeozoic sediments. This period, closing probably at the end of the Cretaceous or the beginning of the Tertiary, was followed by a much longer period of quiescence, corresponding to the deposition of most of the observed sediments. Finally a new cycle of general floor spreading began about 10 m.y. ago *from the same axes* (except for the divergence in the North East Pacific of the same date, brought to light by Vine), thus giving rise to the system of crestal magnetic anomalies and the band of thin sediments.

In order to estimate the duration of the quiet period, the Ewings, for lack of exact knowledge of sedimentation in the Tertiary, suppose that its rate remained the same before and after the discontinuity. They further suppose that the sediments of the crest were accumulated in 10 m.y. In many places on the flanks the sediments are three or four times as thick; the quiet period would have been shorter or, as arguments can be found to show, the sedimentation rate would have been higher at the end of the Tertiary.

Unlike Vine's phenomenon of 25 m.y. ago, the Ewings' seems incapable of any other explanation than a general and more or less simultaneous speed-up on all the ridges about 10 m.y. ago.

In harmony with these ideas, Ewing *et al.* (1968) find that the distribution of Pacific sediments (from the north down to 11°S) is incompatible with any hypothesis calling for a relatively constant rate of floor spreading. Only an intermittent expansion seems acceptable; but obscurities remain. A few details will demonstrate the complexity of the phenomena; no deposit can be found in the North Pacific as thick as those of the Atlantic or Indian oceans, but there are extensive layers, notably the following:

(1) An uppermost acoustically transparent layer, conforming to the topography and visible down to the floor of the trenches. Its thickness, varying from 20 m to 1 km, follows fairly closely the present rate of biological productivity, and shows in particular a strong equatorial maximum. The thickness may be explained, if the layer is made of Tertiary sediments, by appropriate variations in spreading rates, productivity and depth of solution of the carbonates, providing that the expansion followed the equator closely; however the Clipperton fracture makes an angle of some 15° with the axis of the equatorial sediments.

(2) A layer of turbidites generally contemporaneous with the preceding layer and sometimes mingled with it. The superficial sediments near the Aleutians trench and perhaps the Kuriles trench have been attributed to them. Such an attribution implies the absence of the corresponding island arcs at the time, for they would have retained the continental sediments. The expansion was then occurring parallel to the great fossil fractures; the arcs would be due to the recent expansion.

(3) In the western half, a highly stratified layer, thus acoustically opaque, becoming thinner towards the south. It coincides roughly with the assumed limits of Darwin rise, and may be thought to stem from its erosion during the period of maximal volcanic activity, i.e. 60 to 100 m.y. ago, according to Menard (this does not imply that the Darwin rise played the role of an emissive ridge). This type of sediment extends as far as the axes of the Japanese and Marianas trenches, but has not reached the Kuriles trench, towards which the current spreading is driving it.

It will be seen that the consideration of sediments is a delicate matter. At the end of Chapter VII another argument in favour of intermittent spreading will be met with. Ewing and Ewing further cite the case of the Gulf of Aden for which Laughton in 1966 estimated the dates of extension and the thickness of sediments; but it is hard to say whether the quiet periods agree; in the Red Sea, Knott, Brence and Chase assign an episode of tectonic calm to the Pliocene.

6. Overall Pattern of the Anomalies. Their Enumeration and Dating

In four simultaneous articles: A (Pitman *et al.*, 1968); B (Dickson *et al.*, 1968);

C (Le Pichon and Heirtzler, 1968); D (Heirtzler *et al.*, 1968), the Lamont Geological Observatory team attempted to correlate the anomalies observed in the various oceans by identifying as well as possible (fairly well in general) a selection of them numbered from 1 on the crest to 32 on the edge of the pattern without prejudice as to their absolute ages.

Article A for example demonstrates a qualitatively excellent correlation between the North Pacific anomalies along the northern edge of the Mendocino fault and those of the South Pacific as far as the New Zealand plateau (Figure 48). Spreading rates

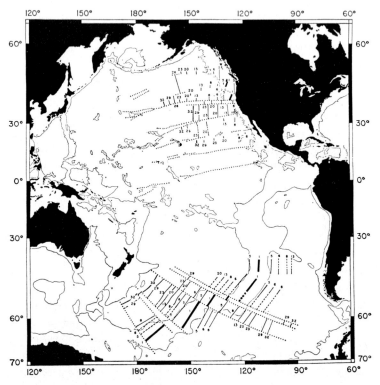

Fig. 48.   Magnetic anomalies in the Pacific; 2000 fathoms contour indicated. Heavy lines locate characteristic anomalies, dashed where the correlation is uncertain. Fracture zones dotted. (Pitman *et al.*, 1968.)

vary along the crest, particularly where fractures traverse it. But a more serious difficulty in assigning an age scale to the anomalies beyond the isotopic one is that the ratio of the rates of the North Pacific to the South Pacific varies with time. In order to adapt it to the South Pacific the authors modify the northern scale, which was already a composite one allowing for the evolution of the ridge.

Article B shows that South Atlantic profiles are internally coherent except for offsets by fractures, and very similar to those of the North Pacific. In the west anomaly 31 coincides roughly with the edge of the Rio Grande rise, in the east anomaly 21 approaches the Walvis rise. Beyond the rises the respective basins of Argentina and the Cape show magnetically quiet zones, as pointed out in Chapter II, succeeded again by more or less clear lineations.

Article C, devoted to bringing out the structural complexity of the Indian Ocean, has already been made use of. In addition to the correlated anomalies of the ridges, it points out on either side of the Ninetyeast rise the possible presence of two ancient east–west anomalies fairly close to the Indonesian trench. The agreement of the crestal anomalies with Pitman and Heirtzler's scale is passable. Spreading rates are 1.5 cm/y. on the Carlsberg ridge, 2.2 cm/y. on the south-eastern branch just south of the Rodriguez fracture, 3.3 cm/y. between Australia and the plateau of the Kerguelens; the slope and relief of the flanks fall off in the same order (cf. Chapter VI); the last section resembles the South Pacific ridge in its topographic calm.

Much greater difficulties occur with the flank anomalies. Interruptions or slow-downs in the spreading rate appear probable for the end of the Oligocene or the

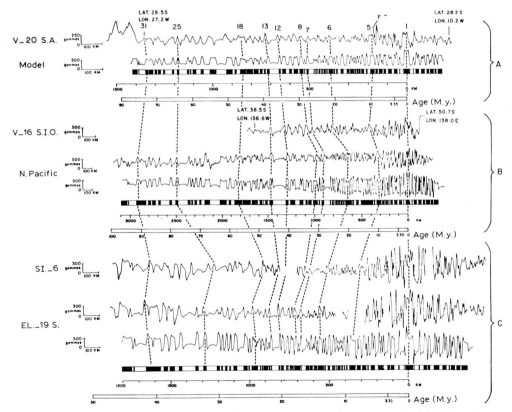

Fig. 49. Selected magnetic profiles from different oceans (S.A.: South Atlantic; S.I.O.: Southern Indian Ocean; SI.-6, EL.-19S.: South Pacific). Below each observed profile is the theoretical one deduced from alternating blocks 2 km thick of normal *(black)* and reversed *(white)* magnetization. The age scale below each model assumes 3.35 m.y. for the end of the Gilbert epoch. Dashed lines connect similarly numbered anomalies. (Heirtzler *et al.*, 1968.)

beginning of the Miocene. They seem to differ somewhat from the general interruption of 30 to 40 m.y. ago which was deduced by Ewing and Ewing (1967) from thicknesses of sediments and which is supposed to have ended 10 m.y. ago at about anomaly 5; Le Pichon and Heirtzler (1968) do not think that this last interruption could have lasted more than 10 m.y. in the south-western Indian Ocean, or more than 15 to 20 m.y. in the center and north. Even so, errors of 20 m.y. would be possible in any reversal scale which assumed a constant rate on a particular ridge.

Article D attempts the synthesis of the other three with a view to establishing an

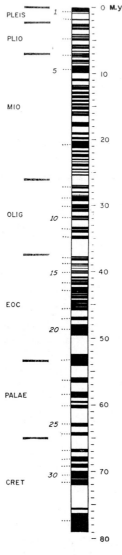

Fig. 50.   Lamont geomagnetic scale, *Left to right*, 'Phanerozoic' scale of anomalies, anomaly number, magnetic polarity (normal in black), age in m.y. (Heirtzler *et al.*, 1968.)

age scale. Figure 49 puts together some magnetic profiles observed in the North and South Pacific, the southern Indian Ocean, and the South Pacific, which represent fairly well the whole of the anomalies of each region, together with the models deducible from the hypothesis of Vine and Matthews. Three age scales are shown; they have been constrained to agree at 3.35 m.y. for the beginning of the Gauss epoch. The correspondence among the three ages of an anomaly with a given number is not linear, therefore it is necessary to make a choice. The Indian Ocean offers too short a scale. The South Pacific scale seems to be in contradiction with 3 observations, the most precise of which concerns a Miocene core taken in anomaly 6, giving it an age of 14 m.y., considered to be too small, while the North Pacific and Atlantic scales give for this core concordant ages (of 20 to 22 m.y.). Since the North Pacific scale applies to a complex crest, the authors finally chose the South Atlantic scale as the definitive reference; it is applicable in the North Atlantic (Reykjanes ridge) as well. An age of 80 m.y. is assigned to anomaly 32, as against 73 m.y. according to Vine (1966). It may be compared in Figure 50 with the *Phanerozoic* scale of the Geological Society of London (Anonymous, 1964).

The preceding scale, which we shall call the Lamont scale, comprises 171 reversals. Most intervals of polarity are briefer than 0.5 m.y. (Figure 51). The average frequency of reversal, fairly constant up until 50 m.y. ago, has subsequently increased to a twofold value at 40 m.y., which has been maintained to the present (around 3/m.y.). It will be recalled that Vine assumed an increase in frequency 25 m.y. ago; such an increase is not confirmed.

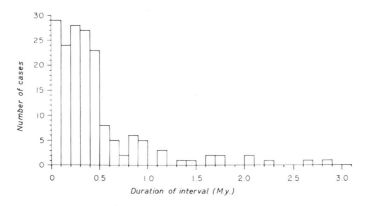

Fig. 51.   Histogram of duration of intervals of polarity. (Heirtzler *et al.*, 1968.)

The Kaena event and a supplementary one subdividing the Olduvai are seen on some profiles only and are not included in the scale. The Gilsa event does not appear. Those of Cochiti and Nunivak, apparently introduced after article D was written, are difficult to place in the scale.

This article's importance lies primarily in the extent in space and time which it covers. The data brought together in it do not contradict the halt in spreading postulated by Ewing and Ewing (1967) on condition that it occurred everywhere at the same time, which does not seem very likely. Nevertheless Le Pichon (1968) takes it into account to modify the Lamont scale by giving anomaly 32 an age of 60 m.y. (Lower Paleocene) instead of 77 m.y., and by assuming that a 10 m.y. halt in spreading began with anomaly 5. Expansion, slowed down by the time of the Oligocene, is thought of as stopping in the Miocene. At the same epoch the trenches along the coast of North America were covered over or filled in; moreover the mid-Atlantic ridge was fractured in the part lying to the north of the Walvis-Rio Grande junction. The links between these three phenomena are still to be established.

According to Le Pichon, nothing that is known about core sample ages or sediment distributions opposes the adoption of his scale. The maximum departure from the Lamont scale is 17 m.y.

CHAPTER V

CONTINENTAL DRIFT

1. Sea Floor Spreading, Drift and Paleomagnetism

The effects of sea floor spreading on the neighboring continents has already been alluded to. The long-standing theory of continental drift was given a direct stimulus when it became known. Only what is indispensable in this theory will be recalled, confined as far as possible to the geophysical aspects.

Sea floor spreading as inferred from magnetic anomalies cannot directly explain phenomena prior to the Cretaceous, but may help in understanding them, so that we shall sometimes go beyond the period covered by the Lamont scale.

Views on the point of departure of drifting are not unanimous. Sometimes it is supposed with Wegener that the continents were united in the Paleozoic in a single Pangaea, sometimes that they formed two groups of comparable area, Laurasia (North America, Europe, Asia less India) and Gondwana (South America, Africa, India, Australia, Antarctica). The exact boundaries of these two domains are disputable. Opinions are equally divided on the date of the splitting-up of the original continent or continents.

The best proofs of continental drift are now those derived from the floor spreading itself or from such of its direct effects as the absence of old sediments in the oceans. The earliest object of the theory was to bring together elements now widely separated which are paleontologically related. We have not the competence necessary for the discussion of such arguments. In general, they assume that in the past the climatic zones were disposed in a manner similar to the present about poles placed elsewhere. In particular the most characteristic and most used climatic indicators are the traces of glaciations considered to result from a situation near the pole. But even these indicators are debatable; Wegener assumed that the opening of the Atlantic was not completed until the Pleistocene, using as argument the extent of the Quaternary glaciations, and although these are still poorly explained, his idea is now considered to be incorrect. On the other hand, the testimony of Permian and Carboniferous glaciations would be a very strong argument for the existence of Gondwanaland.

The facility with which we have turned from the drift of continents relative to one another to their absolute displacement with respect to the pole will have been noted. These two questions are also involved in the paleomagnetic approach to the problem, the sole source of new arguments since Wegener's time. The principle of this approach is very briefly recalled; the magnetic dip and the declination at a given epoch are provided by the magnetization of certain rocks, eruptive or sedimentary, which were cooled or laid down at this epoch. From them the magnetic pole of the epoch is easily deduced; on the average, as we have seen, it coincides with the geographic pole. Although the layers may have been folded it is possible to allow for this to a certain degree.

Using the paleomagnetic results relating to a single continent, South America for example, it is possible to find the path of the north pole with respect to it. (The very great practical difficulties, which have only been taken in hand seriously in the last few years, are passed over.) In the same way, the pole path with respect to Africa is found. If the two continents are made to fit together, as Wegener did, the Paleozoic portions of the two pole paths coincide. The paths diverge in the Mesozoic. Separation would thus have begun in the Permian (Creer, 1965).

Creer applies similar methods to Gondwanaland. According to his tentative conclusions this continent began to break up in the Permo-Triassic. During the Mesozoic, of greater interest here, Australia and India moved farther from the pole than did Africa, while South America's motion practically ceased. Other authors do not see the main separation taking place until the Cretaceous. The paleomagnetic deductions evidently deal with periods preceding those of the magnetic anomalies of the oceans. Indeed, Figure 52 from Runcorn (1965) shows that for at least 60 m.y. the pole relative to North America or Europe departed very little from the present geographic pole.

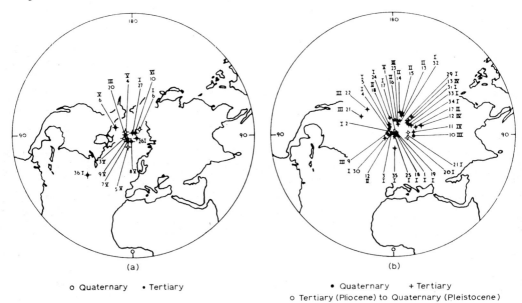

(a)

(b)

o Quaternary   • Tertiary

• Quaternary   + Tertiary
o Tertiary (Pliocene) to Quaternary (Pleistocene)

Fig. 52.   Paleomagnetic pole positions from (a) North American (b) European rocks. (Runcorn, 1965.)

2. Reconstruction of the Continental Blocks

The most striking argument in favor of continental drift has always been the good fit of the South American and African coasts. But Bullard has taken a decisive step forward in looking for the best regrouping of the present continents in an objective manner. Coastlines may vary greatly with small changes in sea level. The true edge of the continent is the continental slope where depth changes from 100 to 200 m to 1000 or 2000 m in the space of a few kilometers. Bullard *et al.* (1965) carried out a mathematical search for the best possible fits of contours corresponding to 100, 500, 1000 and 2000 fathoms depth. They first constituted a block comprising Africa and South America, then a second regrouping North America, Greenland and Europe, and finally they brought these two blocks together (we shall see that the fit is less good in this case).

The Earth may be taken as spherical if the distortions involved, of the order of 200 m, are negligible. Two identical contours on a sphere may be made to coincide by one rotation. Bullard defined the contours to be brought together by a series of points; starting from an approximate center of rotation, he used the method of least squares to find the best value of the angle of rotation corresponding to this center; this value is not perfect and leaves a sum of residuals squared. The center of rotation is then varied systematically in latitude and longitude to reduce the sum to a minimum.

For the fit of South America to Africa best results are obtained with the 500 fathom contours, or about 900 m. It is necessary however to leave out the Niger delta, which is Tertiary, and the Walvis rise, probably Tertiary. The root-mean-square misfit of the two contours, from the mouth of the Amazon to the Cape of Good Hope then becomes 88 km.

This reconstruction is confirmed by the isotopic datings of Hurley *et al.* (1967), confined to western Africa and northern Brazil. In western Africa, Ghana, the Ivory Coast and regions farther west generally yield ages near 2000 m.y. (Eburnean orogeny) by the potassium-argon and rubidium-strontium methods; eastern Dahomey, Nigeria and regions to the east, ages of the order of 500 to 650 m.y. (Pan-African orogeny). The boundary being well-marked, 150 age determinations were made by the two methods in the neighborhood of the point at which it would cross the Brazilian coast in Bullard's reconstruction. A good correspondence is seen (Figure 53) between the date of the Eburnean orogeny and that of the formations of the Guiana shield and of the Brazilian coast just south of the Amazon, but these formations are surrounded by rocks of about 500 m.y. On the boundary, the ages obtained for the micas by the potassium-argon method run from 410 to 640 m.y. while the rubidium-strontium ages, which concern the whole bulk of the samples, are again in the range of 2000 m.y. Farther east the rubidium-strontium ages fall in their turn to 665 m.y. An even better correspondence is established between the Pan-African orogeny and the Caririan orogeny of Brazil, where in both cases ages around 500 m.y. are found from potassium-argon and around 640 m.y. from rubidium-strontium.

Coming now to the North Atlantic, Bullard leaves Greenland separated from Ellesmere Island for want of knowledge of the drift to be assumed in this region. He neglects Iceland, as Tertiary and Quaternary, and the Iceland-Greenland and Iceland Faroes rises, supposed to be Tertiary, like the mid-Atlantic rise. He retains Rockall Bank, Tertiary, which fills in a gap between Greenland and Iceland, assuming it to

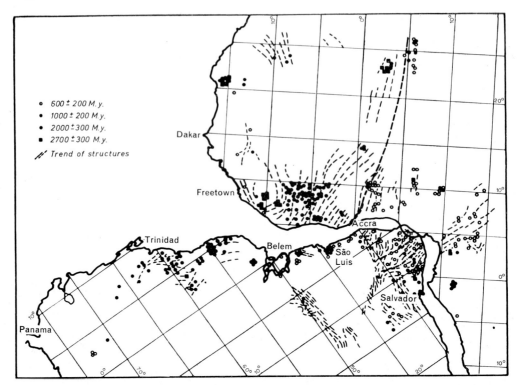

Fig. 53. Fit of Africa and South America according to the reconstruction of Bullard *et al.* (1965). In western Africa the Eburnean province of 2000 m.y. age *(full circles)* adjoins the Pan-African province of 500 m.y. *(open circles)*; their frontier *(heavy dashed line)* would enter Brazil about Sao Luis. Brazilian age determinations show the same provinces as the west African. There may be a similar correlation between western Africa and eastern Brazil north of Salvador. (Hurley *et al.*, 1967.)

constitute an intrusion in an older unknown mass. The final result is almost as good as for the South Atlantic.

Some variants have been suggested, however, for this reconstruction. Thus Wilson points out that in the South Atlantic the Walvis and Rio Grande rises end at the corresponding points of the two coasts. Strong reservations must be taken as to his interpretation of this (that these ridges are constituted from volcanic material issuing from the mid-Atlantic fissure and carried along by the expansion). He then uses the rises of the North Atlantic as a guide to a reconstruction fairly close to the foregoing (aside from the separation of Greenland and Ellesmere Island). Harland and other geologists prefer to place Spitzbergen as Wegener had done, that is to say farther south than Bullard. Harland's solution gets rid of the gap filled by Rockall Bank, but opens up two others to the east and to the southwest of Greenland. According to him, homologous structures are better brought together.

Hitherto, the North and South Atlantic have been considered separately, which amounts to adopting the hypothesis of two independent blocks coming from Laurasia and Gondwana respectively. If they are brought together, Africa overrides Spain, so that continental deformations must be assumed. The least of these is a rotation of Spain closing up the 500 fathom contours of the Bay of Biscay; the hypothesis that this bay opened in the Mesozoic at the formation of the Pyrenees was put forward by Du Toit in 1937, and since supported by various arguments, so far inconclusive; the most recent is the presence of magnetic anomalies probably due to intrusions (Matthews and Laughton, communication at Zurich, 1967). There also remains a

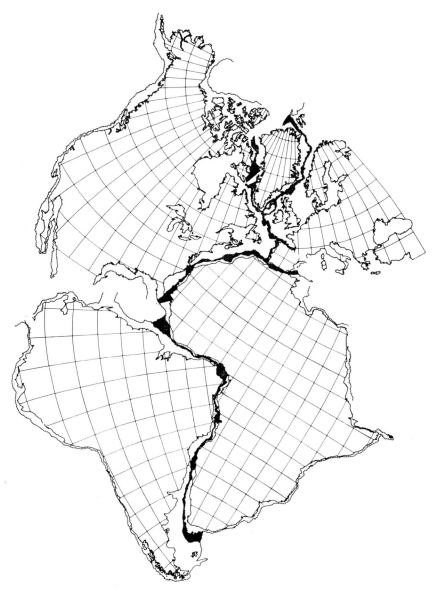

Fig. 54. Continents surrounding the Atlantic reassembled in a single block. *In black:* gaps or overlaps. (Bullard *et al.*, 1965.)

gap of 100 km between Spain and Newfoundland, and the Bahamas shelf must be suppressed. The remaining mean square misfit is 130 km towards the middle of the line.

On the whole this reconstruction (Figure 54) is less convincing than the two preceding. It neglects Central America, Mexico, the Gulf of Mexico, the Caribbean and the Antilles, deformed since the Mesozoic, but whose basement is certainly older. On the other hand, Wilson is pleased to find that a similar reconstruction brings the Hercynian folds of Mauretania close to the Appalachians.

What is the exact significance of this good fit of isobaths? It leads to recognition of the fact that the drifting continental platforms experienced only insignificant amounts of vertical motion, erosion and sedimentation. Fortunately the continental slope is short, so that slight variations may still have occurred. Perhaps the Atlantic is a rare example of the necessary conditions being fulfilled.

3. The Rigid Plate Hypothesis

Probably inspired by Bullard's method for the reconstruction of continents, Morgan in 1967 (Morgan, 1968) on one hand, then McKenzie and Parker (1967) on the other, tried to explain continental drift on the supposition that thrusts stemming from the axis of ridges displaced, *without materially deforming them*, great spherical plates comprising both continental and oceanic surfaces, bounded by seismic lines. The only deformations permitted occur on these lines, that is, the active ridges on one hand where the emitted matter would add to the surface of the plates on either side, and trench regions on the other, as well as regions of folding and overriding where portions of the surface would disappear. The other boundaries of the plates would consist of slip faults. Evidently only highly schematic models will be obtained in this way, but they are most valuable nevertheless.

It seems rather unlikely that a local phenomenon like the injection of dikes in ridge axes or the dip-slip of ridge layers should be able to move a continental block. Hence the phenomenon must be at great depth, with the slipping surfaces lubricated, as Orowan suggests. It is preferable to follow Elsasser (1968) in imagining the rigid plate (lithosphere, or tectosphere as Elsasser calls it) to be carried along by under-lying convective movements, which may be somewhat irregular, with the plate integrating their effects.

In this concept the ridge position is not tied down too closely to a highly localized thermal upwelling which would be difficult to justify. There need only be currents which are on the whole divergent to produce a fissure along a line of weakness of the lithosphere which would immediately be injected. As the currents continue to act, the plates separate and a new fissure is produced in the center of the old because it is the hottest part and therefore once again the weakest. This fissure, and hence the ridge crest which will continue to form by the same process, will not be stationary on the Earth's surface unless the two sides of the initial fissure spread apart sym-metrically; but in all cases the structure of the ridge thus produced and its magnetic anomalies will necessarily by symmetrical with respect to the instantaneous crest. If the initial fissure occurred under a continent, its edges are those of the present continents; the ridge is mid-oceanic.

Of course, all this is schematic. The evolution of the lithosphere must react on the convective mantle; perhaps this takes place in the course of great readjustments like those of the North-east Pacific. It may be asked, besides, what thickness this spherical cap of lithosphere must have in order to transmit the requisite forces over large distances, if not from one side to the other. This point had already been debated in connection with Wegener's theory. Jeffreys had concluded that the compressive strength of ordinary rock would be exceeded for a plate of continental dimensions floating on liquid.

The question was reconsidered by Orowan (1965). On Griffith's theory the strength of a rock depends on the largest crack (often submicroscopic) present in the specimen. A strength on the order of 1000 to 2000 bars like that of granite or basalt corresponds to Griffith cracks of from 1 to 10 μ length, without friction between their walls. The upper part of the crust shows joints of several meters length; and the strength of this crust would fall to a millionth of the above if it were not for friction between their faces. But this is surely operative in the deeper parts. Orowan applies the rules of soil mechanics to the top 10 km of the continental crust, assuming Coulomb's law for the friction. He gets on the order of 6500 bars for the compressive strength (against only 1100 bars for the tensile strength, which perhaps explains the opening of fissures in the continents). In light of these figures, Orowan accepts Wegener's view of the continents' displacement in the oceanic substratum. Our problem is somewhat different, and the calculation open to argument, but the result is encouraging.

Orowan also considers the extreme case of a continent with no cohesive strength. It will still have what Orowan calls 'gravitational strength'; a continent subject to a horizontal strain increases in thickness and thus resists the strain or possibly transmits it. By a simple hydrostatic calculation, Orowan shows that a continent of density 2.8 g/cm^3, thickness 33 km, floating on a fluid of density 3.3 g/cm^3 would have a compressive strength of the order of 700 bars. If the thickness is tripled to that of the lithosphere according to Oliver and Isacks, the strength becomes adequate.

Thus the drift of rigid plates both continental and oceanic does not seem absurd if they are assumed to be sufficiently thick. This leads to difficulties in some regions, such as California, where the earthquakes of the San Andreas system have focal depths under 16 km (Bolt in 1965) and where efforts have been made to compare the fault faces to plates of such small thickness.

4. Kinematic Considerations and First Applications of the Plate Hypothesis

The rigid plate hypothesis permits study of the kinematics of surface movements as soon as sufficient data are available for some of their boundaries, for example the direction of spreading on those portions of boundary consisting of ridges. This expansion may take place obliquely to the crest while still giving rise to anomalies parallel to it. Nothing in the present state of our knowledge however forbids the

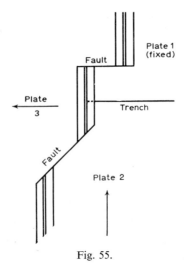

Fig. 55.

assumption that it is always perpendicular to the crest, in spite of the obliquity of the transverse fractures interpreted as transform faults; this obliquity may occur simply because the velocity of one side of the ridge with respect to the other has a component parallel to the axis.

Figure 55 shows an example in plan (for simplicity) taken from Morgan (1968). In it the side of the ridge next to plates 1 and 2 remains fixed, but the *ridge axis drijts* at the half-rate of plate 3. The simultaneous existence of both a normal and an oblique fault is due to the presence of the intervening trench. If two plates are simply separated by a ridge intersected by transverse faults, these latter are necessarily parallel.

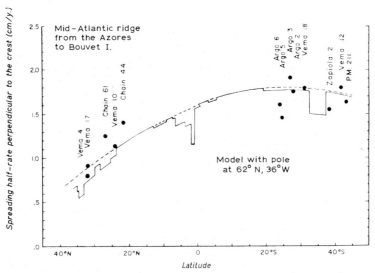

Fig. 56. Comparison between spreading rates deduced from magnetic anomalies and from the separation of two plates; *full line* rate perpendicular to ridge; *dashed*, parallel to the direction of spreading. (Morgan, 1968.)

On the sphere this parallelism condition is replaced by the following: the borders of two plates pushed apart by an intrusion in the axis of a ridge may be restored to coincidence by a small rotation about a certain instantaneous center; the faults offsetting the ridge must then be portions of circles having this instantaneous center as pole. The width of the small fissure, or equivalently the component of the drift rate in the direction of these parallel circles, increases (as the sine of the angular distance from the pole so defined) as far as the corresponding equator, and thereafter decreases. This law of variation has been verified by Morgan (1968) for the mid-Atlantic ridge from the Azores to Bouvet Island (Figure 56) using the mean drift rate for the last 5 m.y., and for the East Pacific rise.

Attention also may be drawn to an additive relation, enunciated by McKenzie and Parker, that obtains between the relative rotations of three plates when these have one common point (where, for example, a fault, a trench and a ridge meet).

How the fundamental hypothesis is applied depends on the geographical choice of the rigid plates and the boundary conditions.

McKenzie and Parker (1967) treat the North Pacific as a moving plate and study its displacement with respect to a stationary plate comprising North America and the north-eastern part of Asia assumed to be rigidly connected in the region of Behring Strait. The boundaries of the mobile plate are: the American coast from the Gulf of California to Alaska, the Aleutians, Kamchatka, the Kuriles, Japan as far as the Fossa Magna, ending in the south in a line of closure, presumably oceanic, but not specified. The two ends of this line are examples of triple points: three island arcs originate at the Japanese extremity, while the San Andreas fault, the East Pacific rise, and the Central American trench following the continent as far as the Gulf of Panama originate from the Californian extremity.

McKenzie and Parker use the direction of first motion of 80 American earthquakes, assuming that the *Vela Uniform* seismographic network has made such determinations thoroughly trustworthy. It is still necessary to resolve the ambiguity of $90°$ inherent in the method (cf. Chapter I), but only one of the possible directions varies in a continuous and systematic manner throughout the length of the land boundary of the plate.

The authors in fact take as conditions at the limits the direction of the San Andreas fault between Parkfield and San Francisco for one end, and the average horizontal displacement during the repeated shocks of the great Kodiak Island earthquake in Alaska in 1964 for the other. These conditions determine a center of rotation located at $50°N$, $85°W$, and from this one can calculate the displacements at other points of the boundary, which may be compared with the first movements of the 80 earthquakes. The agreement is remarkable. Moreover the movements obtained correspond to the tectonic requirements: the great American system of en echelon faults of San Andreas, the Queen Charlotte Islands (Canada) and Fairweather (Alaska) corresponds closely to strike-slip movement. From the place where the seismic zone changes direction, Alaska and the islands overthrust the Pacific at a small angle (around $7°$). The curvature of the Aleutians gradually reestablishes the slip movement;

the authors draw attention to the simultaneous decrease in importance of trenches, andesitic volcanoes and deep-focus earthquakes. These phenomena, and the related one of overthrusting, increase again at the sharp bend approaching Kamchatka.

An evident shortcoming of McKenzie and Parker's work is that it assumes the Pacific rim to be indeformable in spite of the influence of the opening of the Atlantic. Such a defect is unavoidable in any regional solution. The treatment of Morgan (1968) is a good deal more ambitious, since he considers the surface of the entire globe, divided in 20 plates whose size varies from that of the Pacific or of Africa down to that of the Antilles, Iran, or even a small area next to the Juan de Fuca ridge. The African block is separated from the Eurasian block by the Azores-Gibraltar rise considered as a major strike-slip fault.

Generally speaking the boundary conditions used by Morgan are not given by the first motions of earthquakes but by the directions of transverse fractures and by the spreading rates on the ridges or the rate of slip on the San Andreas fault. The data are unfortunately insufficient to permit a complete discussion of the displacement of the 20 plates. Among other results, Morgan, like McKenzie and Parker, finds that the Pacific is no longer moving towards New Guinea and the Philippines as it did when the great fossil fractures were active, but towards the Aleutians and Japan.

5. The Synthesis of Le Pichon

In order to examine the compatibility of drifting on a worldwide scale, Le Pichon (1968) reduced the number of plates to be considered although he had more data at his disposal than Morgan, and he assumed that the expansion was perpendicular to the crests.* The Mercator projection of Figure 57 shows the ridges he considers to be active, and the 31 spreading rates, i.e. half the speeds of separation of the plates, which he uses. The values are given to 0.1 cm/y. and vary from 1 to 6 cm/y. The figure also shows the location of anomaly 5, corresponding to 10 m.y. in the Lamont scale. The markings for the south of the Indian Ocean are rather uncertain.

The centers of the relative rotation of the two sides of an active ridge may be determined by a least squares procedure starting either from the vectors representing the expansion perpendicular to the ridges or from the segments of parallel circles representing the transverse fractures. Le Pichon checked the agreement of the two determinations by moving the pole of the Mercator projection to the center of rotation SP or C_1 corresponding to the South Pacific ridge. In Figure 58 we see how the constituents of this ridge and its transverse fractures become parallel to the two axes of the map.

It may also be verified that any numbered anomaly remains at a distance from the ridge proportional to the sine of its angular distance from the pole, which is constant in the projection used.

The center of rotation SA applicable to the South Atlantic ridge is close to the pole

* At least in the preprint used here, though not in the final article. Results are the same in any case except for the Atlantic.

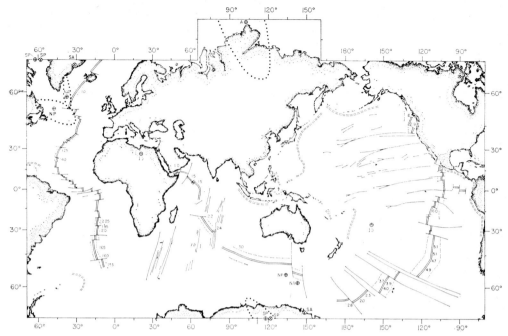

Fig. 57. *Dashed*, anomaly 5 (10 m.y.). Spreading rates in cm/y. Centers of rotation from spreading rates $\otimes$, from azimuth of fractures $\oplus$; NA, North Atlantic; SA, South Atlantic; NP, North Pacific; SP, South Pacific; IO, Indian Ocean; A, Arctic. The ellipses dotted around the centers are the loci of the points $\sigma = 1.25 \ \sigma_{min.}$, and show the convergence of the least squares method. The ellipse around IO is too small to be shown. (Le Pichon, 1968.)

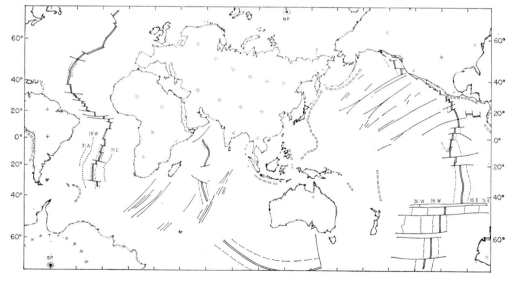

Fig. 58. With SP as pole of the projection (69 °S, 123 °E). Anomalies 5, 10 and 31. (Le Pichon, 1968.)

C_1 and the properties mentioned above remain approximately true for this ridge. The great fossil fractures of the East Pacific clearly belong to a different system; however in places the anomalies numbered 5 are perpendicular to them, which is understandable, since the change in direction of the expansion took place just at the epoch of emission of these anomalies.

The center SA was rather poorly defined by the sea floor spreading south of the Azores. The fractures lying between $30°$N and $8°$S define a centre NA or C_2 2000 km distant (Figure 57) which Le Pichon adopts for the whole Atlantic ridge, not being able to take a relative movement between the two Americas into account.

Although there is a rough agreement between the Atlantic and the Pacific centers of rotation, the Arctic and Indian oceans behave quite differently. Le Pichon considers their ridges to be oblique openings linking the Atlantic and Pacific ones. Having little data from the Azores to the mouth of the Lena, he assumes that the Arctic ridge is separating Eurasia from the Greenland-America block at the rate of 1 cm/y., but the center of rotation A or C_3 is ill-determined from non-parallel fractures.

In order to represent the movement of the Pacific with respect to America, Le Pichon schematizes the complex structures of the American coast as a series of ridges and transform faults running from the Gulf of California to the Aleutians arc. Although the dispersion in the direction of fracture zones bounding the Juan de Fuca ridge reaches $20°$, he derives from them a center of rotation NP or C_4 ($53°$N, $47°$W) lying over 2000 km from McKenzie and Parker's center, but fairly close to Morgan's ($53°$N, $53°$W). Nearly concordant values for the speed of rotation are obtained from the spreading rate according to Vine (2.9 cm/y.), or from slip on the San Andreas fault, estimated at 6 cm/y. by Hamilton and Myers in 1966, and as 300 km since the Upper Miocene by Rusnak, Fisher and Shepard in 1964 from the width of the Gulf of California (Figure 59).

The spreading rates and the fractures of the Indian Ocean remain to be utilized. The fractures, poorly known in places, have various directions, with north-north-east to south-south-west predominating. Le Pichon makes use of the great Owen fracture, and the fractures of the Gulf of Aden defined by Laughton in 1966, although his determination has been criticized subsequently. In this way the center IO or C_5 is obtained for the rotation of the north-eastern part of the Indian Ocean with respect to Antarctica; this center suits the Carlsberg ridge fairly well, but is no longer appropriate to the south of Rodriguez Island.

From the five centers C_1 to C_5 of the rotations corresponding to the opening up of the ridges, the relative rotations of any two adjacent plates, up to a number of six, may be deduced by calculation. Seismic geography suggested to Le Pichon the choice of Eurasian, American, African, Pacific, Indian and Antarctic blocks, as shown in Figure 60, where the Arctic regions are unfortunately missing. The boundaries on which the spreading rates are known are shown as double lines. The dashed curves are subsidiary boundaries which Le Pichon would have wished to incorporate.

As it is, the model allows calculation of the recent relative displacement between adjoining plates at 37 points on the boundaries with a precision which falls off as the

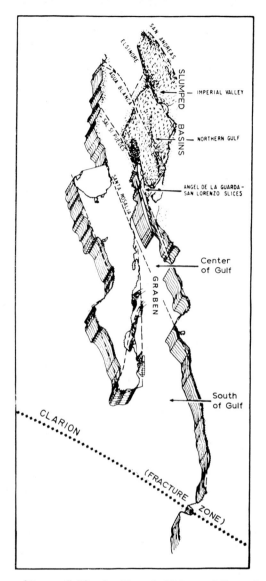

Fig. 59. Block diagram of Lower California. (Rusnak, Fisher and Shepard; 1964, *Marine Geology of the Gulf of California*, Van Andel, Shor, Ed.; Am. Assoc. of Petroleum Geologists.)

number of rotations involved increases (as many as four between the Indian and African blocks). We summarize the principal results:

An extension is indeed obtained for both branches of the Indian Ocean ridge, but the rate is only 1 cm/y. on the south-west branch compared with 3 cm/y. for the south-east.

Shortening at the rate of 8 or 9 cm/y. is found for the island arc region of the Western Pacific, with strike-slip motion to the north of the Fiji Islands, and a similar smaller

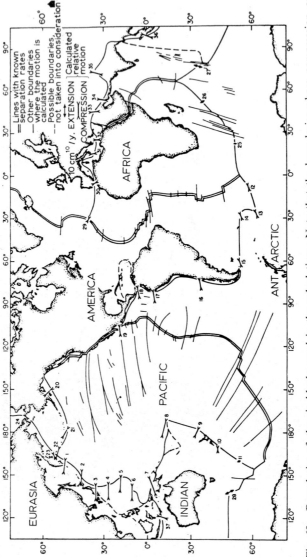

Fig. 60. Boundaries of the six blocks used in the calculations. Note that the boundaries where compression or slip exceeds about 2 cm/y. give rise to most of the Earth's seismic activity. (Le Pichon, 1968.)

motion (5 to 6 cm/y.) to the west of the Aleutians, thus agreeing with McKenzie and Parker's result. Over a much shorter length, the South Sandwich arc also shows as a region of compression bordered by strike-slip motions, smaller, however (2 to 4 cm/y.).

The Chilean coast shows shortening movements at the rate of 5 to 6 cm/y. They extend to the precontinental rise of Panama, which appears to be a filled-in trench, perhaps marking (like a similar trench east of the Juan de Fuca ridge), the location of the ancient island arc system of the East Pacific. However, this recent compression of a hidden trench seems ill-established because of the omission of the Antilles arc and the zone of the Galapagos.

Finally, the region of Alpine foldings, on the site of the ancient Tethys Sea, corresponds to a shortening rate which increases from 2 cm/y. near the Azores to 6 cm/y. near Java, and is thus less than on the oceanic trenches.

On the whole this purely kinematic model is undeniably successful (Le Pichon considers the solidarity of the two Americas to be its main shortcoming). But the model should not be taken to imply that the oceanic ridges determine the tectonics of the whole Earth. The regions of folding are often earlier than the ridges, and their location may have influenced the position of the latter.

6. Ancient Sea Floor Spreading

The rigid plate hypothesis seems to have been applicable in the distant past, judging from the aspect of the transverse fracture zones. In principle, the hypothesis allows the reconstitution of the continents and their adjacent basins as they were at the time of the emission of a given anomaly, by moving the plates so as to bring together the two appropriately numbered curves on either side of a ridge. This is of greater interest, of course, if the epoch is known. The time scale of Le Pichon, which differs from the Lamont scale in assuming discontinuities in spreading rates, has already been mentioned (Chapter IV).

Figure 61 shows the positions of the continents at the time of anomaly 5, i.e. at the end of the Miocene, with respect to Antarctica arbitrarily taken as fixed. The Tethys has closed in since that epoch on the order of 200 to 500 km.

The same method may be applied to anomaly 31, dating from 60 m.y. (Paleocene), as to anomaly 5. Such an extrapolation is rather bold because this anomaly has not everywhere been recognized. We shall limit ourselves to the least conjectural result, concerning the Atlantic (Figure 62).

Bringing the corresponding curves number 31 together amounts practically to the suppression of the present ridge surface, the lateral basins coming into contact. As Morgan had already noted, the lines of fracture so obtained, which may be regarded as the lines of the earlier displacement, differ considerably from the present ones; this probably reflects the change in the pattern of spreading. These ancient lines do correspond well with known topographical features; the northern edge of the plateau of the Falklands and the fault zone which follows, the escarpments to the north of

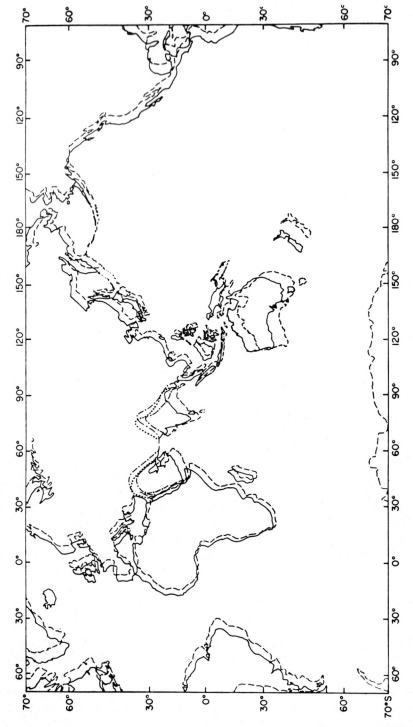

Fig. 61.   *Dashed lines*, positions of the continents with respect to Antarctica at the time of anomaly 5 ($\sim$ 10 m.y.); *continuous lines*, present positions. (Le Pichon, 1968.)

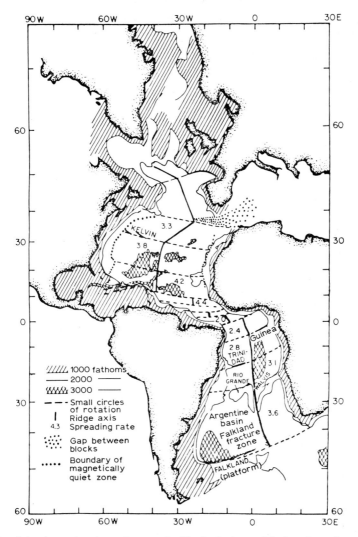

Fig. 62. The Atlantic at the time of anomaly 31. *Dashed*, possible location of great Mesozoic fractures, got by assuming that pre-Cenozoic drift was equivalent to a single rotation for the two Americas at the average rate between −120 and −70 m.y. Bathymetry not shown for the Caribbean and the Gulf of Mexico. Trends of the main topographical features agree well with the directions given by the rotation. (Le Pichon, 1968.)

the Walvis and Rio Grande rises, the rise on which Trinidade Island is located east of Brazil, the Guinea rise and the Kelvin seamount chain.

The Miocene episode makes it difficult to assign an average rate for spreading since the Paleocene, but this can be done for the earlier spreading, dated from the opening up of the Atlantic. It should be mentioned however, that to arrive at the state of the North Atlantic for this moment, Le Pichon does not fix the coincidence at the 500 fathom contours used by Bullard. He uses the zones of magnetic quiet

(Chapter II), which are more symmetric about the ridge, although not fully so. Unlike Heirtzler and Hayes, who look upon these zones as regions of Triassic or Permian spreading during a long period free of reversals, Le Pichon sees them as continental margins which subsided in the Lower Cretaceous, at which time active spreading started, in his view. But did formation of oceanic crust, 'oceanization' in Beloussov's terminology, take place?

If it is supposed that the Mesozoic spreading started at -120 m.y. and ended at -70 m.y. in the situation depicted in Figure 62, the Mesozoic rates there given are found. They show a discontinuity at the equator, which Le Pichon attributes to the creation of the Antilles; it is to be recalled that the two Americas form a single block in the model. The gap between Europe and Africa may have been closed during the Paleocene by thrust from the median ridge of the Labrador Sea; this movement too is incompatible with Figure 62, which assumes Greenland to be already separated from North America. Le Pichon is concerned by the fact that the South Atlantic ridge is not in a median position in the figure. He assumes that the crest shifted eastward at the beginning of the Mesozoic; this point remains obscure.

Without attributing too much exactitude to these reconstructions at remote epochs based on the hypothesis of rigid blocks, it must be recognized that they at least furnish a plausible pattern for continental drift during those geological periods which left magnetic evidence.

CHAPTER VI

STUDY OF THE OCEANIC CRUST
BY SEISMIC REFRACTION AND GRAVIMETRY

1. Seismic Refraction. Crustal Structure

The phenomenon of sea floor spreading cannot be understood without a knowledge of the structure of the crust and the underlying mantle. The large number of seismic profiles carried out in the last few years has considerably changed the picture gained from the study of near earthquakes more than 20 y. ago. At that time the continental crust was imagined as a series of geological layers defined by the average velocities of seismic waves in them. Beneath a highly variable sedimentary layer a so-called granitic layer was recognized, separated from the next layer, termed basaltic or gabbroic, by the doubtful Conrad discontinuity. The crest was separated from the upper mantle, thought to be composed of peridotites rich in olivine $(Mg, Fe)_2SiO_4$ by the Mohorovicic discontinuity, or Moho. Just below the Moho, the velocity of P waves was 8.1 km/s, subsequently increasing with depth.

All this is hardly more than a rough outline; the boundaries of the layers (and even the Moho) are rarely well-defined, and in any case are neither plane nor horizontal. The velocities vary within the layers. Finally, they do not increase constantly with depth, for the existence in the mantle of a 'low-velocity layer' (or possibly even two such layers) seems to be established (Chapter VIII). Unfortunately, seismic velocities do not provide any very exact indication of the nature of the rock; they may remain much the same despite major changes in rock composition, in exceptional cases even having identical values for rocks which are petrographically very different.

Particular interest attaches to geographic variations in the depth of the Moho, whose extreme values are around 10 and 70 km, and to the variation of seismic velocities below it. For the P waves velocities of up to 8.5 km/s have been found for the center of the United States; more importantly, mantle with low velocity (under 7.8 km/s) has been discovered east of the Rockies where the ancient crest of the East Pacific is supposed to have been. Such *anomalous mantle*, with velocities as low as 7.3 km/s was found by Tryggvason in 1964 under Iceland, at 8 km depth under the Ivrea zone during the experiments in the Alps of 1956, 1958 and 1960 (Labrouste *et al.*, 1968), and in several other tectonically similar regions.

Seismic exploration of the oceans, begun by Ewing about 1935, being easier than that of the continents, has progressed much more rapidly. The results are quite coherent if the ridge crests, trenches and continental shelves, which form a relatively small area, are eliminated. Non-consolidated sediments are found first in highly variable thicknesses, which are studied by means of sounders giving continuous profiles, and in which the *P*-wave velocity, starting from the velocity of sound in water (1.5 km/s), increases with the compaction of the sediment, but remains of the order of 2 km/s. Under the sediments appears the 'basement', often called 'layer 2'; its thickness is about 2 km and the velocity from 4 to 6 km/s. It is probably heterogeneous, essentially composed of intrusive rocks and volcanic outflows incorporating consolidated sediments, which remain unrecognized where they are overlain by higher velocity materials. Numerous dredge hauls of basaltic rocks and photographs of lava flows support this interpretation. It is also to be recalled that trial drilling for the Mohole encountered basalt a few hundreds of meters below the ocean floor.

Below these two layers, that is, beyond 6 to 8 km below sea level, the results become remarkably uniform. In all the oceans is found a layer of average velocity 6.7 ± 0.3 km/s which is called the 'oceanic layer' or 'layer 3'. At about 10 to 12 km depth the velocity rises to 8.1 ± 0.3 km/s in crossing the oceanic Moho; this is the value found below the continental Moho at much greater and more variable depth. Nevertheless the continuity of the two Mohos seems to be established by several profiles perpendicular to coasts (Figure 63).

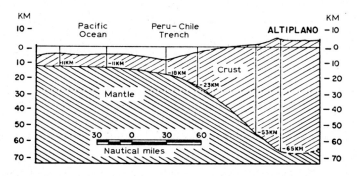

Fig. 63.   Profile of the continental margin off the Andes. Continental data from *Carnegie Institution of Washington*, oceanic from *Scripps Institution*. (From *American Bulletin for IGY*.)

2. Seismic Study of the Ridges

After this review of general results we can take up study of the ridges. Figure 64 shows refraction profiles of Lamont Observatory (Le Pichon, 1966). They concern the south of Iceland, the Azores, the equator, and are not sufficiently numerous for proper statistics. The authors admit that the bathymetry is not always good, but the continuity of the profiles and the fact that the shots were logged in both directions enable a passable determination of the structure to be made.

In the figure the region bounded by the 3.5 km isobath is indicated as the axial zone. There the sub-crustal wave velocity is around 7.3 km/s, actually from 7.1 to 7.6 km/s; that is, anomalous mantle is present. Between 3.5 and 5 km of ocean depth (without exception from 3.9 km downward), the normal velocity of around 8.1 km/s is observed.

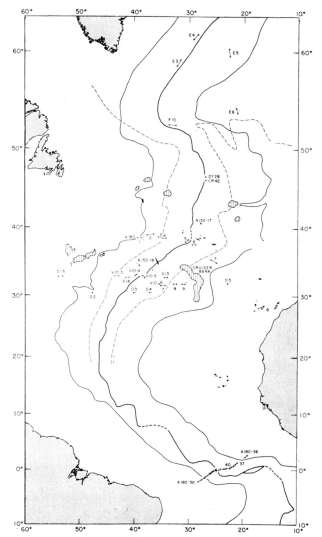

Fig. 64. Location of seismic refraction soundings carried out by *Lamont Geological Observatory*. *Solid lines*, axis and edges of the ridge; *dashed*, approximate limit of the axial seismic belt. (Le Pichon *et al.*, 1965.)

The existence of anomalous mantle under the crest of ridges is a general result, characteristic of their seismic structure; the less precise study of certain seismic surface waves (Love waves) has shown that the drop in velocity of P waves is accompanied by a similar drop in velocity of S waves.

If the mantle is considered to be anomalous when the P-velocity falls below 7.9 km/s, notable examples of this property are found below the East Pacific rise (7.5 km/s), the Carlsberg ridge (7.8), the median trench of the Red Sea (7.1), the Gulf of California (7.9) and under the vanished ridge of the Labrador Sea. It is also found under such island arcs as Japan (7.7; report to the Upper Mantle Committee, 1967) or the New Hebrides (7.4; Dubois, 1966), and even in regions whose present structure is quite unlike that of the ridges and arcs, but where an older extension may be suspected, for example a part of the western Mediterranean (7.7; Knopoff, 1967) or the Irish Sea (7.4; Matthews and Laughton, communication at Zurich, 1967).

Cook (1962) pointed out that the practically linear curve derived by Gutenberg for the velocity of *P* waves as a function of depth in the upper mantle would give a velocity of 7.4 km/s at the surface if extrapolated (Figure 65). Thus anomalous mantle would correspond to an influx, without differentiation, of materials situated under the low-velocity layer. But this is an over-simplification from a petrographic standpoint.

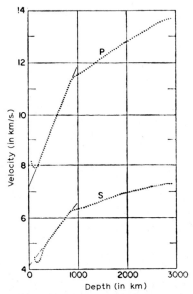

Fig. 65.   Velocities of *P* and *S* waves in the mantle according to Gutenberg, extrapolated to the surface. (Cook, 1962.)

Figure 66 (Le Pichon, 1966) confronts four Atlantic profiles with one from the East Pacific rise obtained by Raitt in 1956 and redone by Menard in 1960. Two observations of normal mantle near the crest will be noted in profile V 10. Whitmarsh (1968) comments that these high values, and others like them, correspond to seismic profiles perpendicular to the axis of the ridge (or, in one case, perpendicular to the axis of the Gulf of Aden). He considers them to be a result of a pronounced anisotropy in the anomalous mantle, diminishing at great distances from the axis while the velocities increase. We shall soon see the evidence brought forward by Hess (1962) for this kind of anisotropy. It should be said that on the other hand, a velocity of 6.8 km/s is found at one point of the anomalous mantle, which suggests prudence in these interpretations.

It will also be seen that the 'oceanic layer' disappears abruptly towards the axial

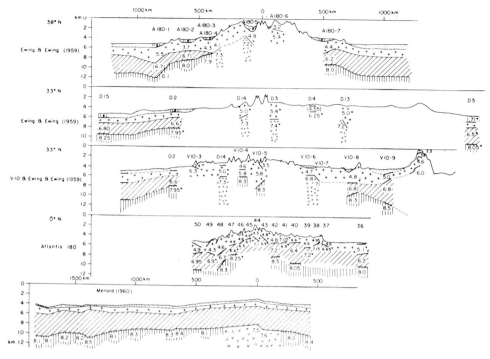

Fig. 66. Crustal sections across the mid-Atlantic ridge (four upper profiles) and the East Pacific rise (below), *Dotted*, unconsolidated sediments; *crosses*, basement; *oblique hatching*, oceanic layer; *vertical hatching*, normal mantle. (Le Pichon *et al.*, 1965.)

region in the first three profiles. It is similarly absent in three seismic surveys on the Reykjanes ridge, including one fairly distant from the crest (Figures 64 and 67). By contrast the 'basement' is always present in the Atlantic profiles, from the crest to the abyssal plains, with the same roughness but greater thickness and velocity in the axial zone (Le Pichon *et al.*, 1965).

Despite this slight thickening of the basement, the crust becomes rather thinner in the axial region. This is contrary to the usual idea of isostasy, whereby all relief should be compensated by a root in the Moho, and also contrary to what occurs on purely volcanic rises.

The equatorial profile, which cuts the Atlantic ridge in its narrowest part, just north of the transverse fractures, and particularly the East Pacific profile, are different. The oceanic layer does not disappear below the crest; it remains at something like 4 km under the Pacific ridge compared with 5 km under the basins. The basement thickness and velocity are not significantly different from the flank values. The crust is normal ocean crust, but slightly arched.

It is often claimed that ridges evolve from the Pacific type, without median valley and with low relief in the crestal zone, towards the Atlantic type. But Van Andel and Bowin (1968) point out that broken relief and sediment-filled valleys are found all along the Atlantic ridge, whose character must thus have remained the same from

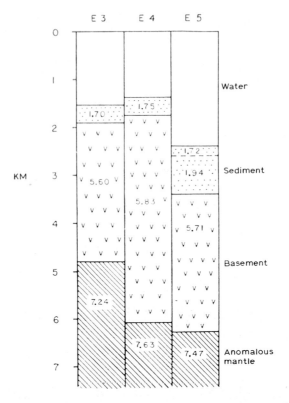

Fig. 67.   Results from three refraction profiles of Ewing and Ewing (1959) (the most northerly in
Figure 64) in connection with the Reykjanes ridge.

the beginning. Ridge differences stem, in part at least, from differences in the spreading
rate. Table I, due to Menard (1967b), shows its relation to the relief and basement
thickness. Note that the total thickness of the crust is not considered.

TABLE I

Region or ridge	Spreading half-rate cm/y.	Central rift	Flank relief in crestal zone	Average thickness basement, km	Lava production km³/km m.y.
Iceland	0.3 to 1.0	yes	mountains	3.5	25
Atlantic					
30° to 40°N	1.0	yes	mountains	2.9	29
Red Sea	1.0	yes		3.3	33
Reykjanes	1.0	no	mountains	3.3	33
Gorda (crest)	1.0	yes	mountains		
Carlsberg	1.5	yes	mountains		
South Atlantic	1.5	yes	mountains		
Juan de Fuca	2.9	no	hills	1.5	
Gorda (flank)	3.0		hills	1.0	30
East Pacific 50°S	4.5	no	hills	1.0	45

Rates up to 1 cm/y. are associated with a thick basement and a median valley (except for the Reykjanes ridge), while from 2.9 cm/y. up the basement is thin. From this it follows that the rate of production of lava per kilometer of ridge, calculated from the thickness of basement and the spreading rate (neglecting what may occur in the sediments or the oceanic layer), does not vary greatly. Lava production from all the ridges would amount to 5 to 6 km³/y., while 1.8 km³/y. from the Earth's origin would have sufficed to produce the volume of the continents. Finally, low rates give rise to strong relief, possibly because they allow more time for volcanism and the formation of normal faults.

Menard has tested these ideas in the South Pacific, where the ridge has a relatively smooth crest, the middle flanks have mountainous bands, and the outer flanks are again smooth. Now here the spreading rate has varied, the three respective values being 4, 1.5 and 4 cm/y.

3. Dredgings of Igneous Rocks

We have been cautious in identifying the oceanic layers. A predominance of basic igneous rocks has been assumed for the basement. In the upper mantle, oceanic or continental, the presence of peridotites, with or without pyroxenes, may be envisaged. Lastly according to Hess (1962) the oceanic layer, with P-wave velocities between 6.0 and 6.9 km/s, could be derived from the peridotites of the mantle serpentinized to about 70%, the percentage corresponding to the average P-wave velocity of 6.7 km/s. On the other hand Cann (1968) considers the rocks to be amphibolites resulting from the metamorphization of basement basalts (Chapter IX).

Dredgings are not very conclusive as to these two hypotheses. Apart from some erratic rocks transported by icebergs, the rocks dredged up on the ridges are basic or ultrabasic, more or less hydrated or metamorphosed. Most are basalts, often vesicular or partially glassy, produced by under-water volcanism; some are fresh but others are altered, particularly by hydrothermal means. Many samples appear to have been chilled abruptly (pillow lavas). It is these basalts characteristic of depths greater than about 1 km (tholeiitic basalts), less alkaline than the olivine basalts of the islands or seamounts, which are probably responsible for the alternating magnetic anomalies, at least when fresh, for hydrothermal changes decompose the magnetic oxides.

In accordance with Cann's ideas, basic rocks and basaltic tuffs metamorphosed to the greenschist facies, have been brought up from the east flank of the Atlantic valley (Van Andel and Bowin, 1968).

However, besides gabbros and basic breccias, more or less serpentinized peridotites metamorphosed by crushing have been recovered from the main transverse fractures and normal faults of the median valley. The St. Paul Rocks, lying on the Atlantic ridge just north of the equatorial Chain fracture, are composed of serpentinized dunite; a deep haul (9400–9700 m) on the coastal flank of the Tonga trench (Fisher and Engel, 1968) recovered 10 kg of basalt and 27 kg of dunite, of which 10 kg showed various changes into serpentine. This would appear to favor the Hess hypothesis.

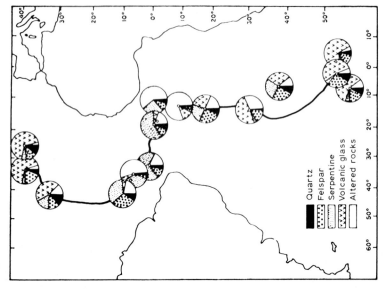

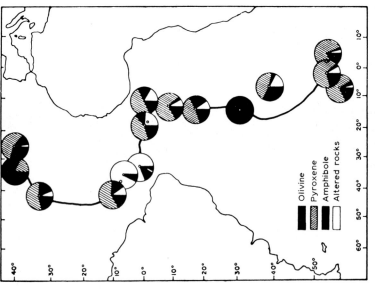

Fig. 68. Mineralogical composition of dredged sands. *Left*, heavy minerals; *right*, light. Samples with less than 300 grains of heavy minerals omitted. (Fox and Heezen, 1965.)

Dredgings of sands on the Atlantic ridge (Fox and Heezen, 1965) are equally inconclusive. On the crest itself, usually only a few meters of sand are found, at times with a marine constituent (tests of foraminifers), and always with the decomposed minerals of a tholeiitic basalt, sometimes accompanied by volcanic glass (Figure 68). The equatorial region, with numerous fractures, is noteworthy for the proportion of hydrated materials, serpentine and uralite, derived from peridotites and pyroxenes respectively. But amphibole is abundant.

To sum up, the abundance of hydrated ultrabasic rocks on one hand and amphibolites on the other is undoubted, but what can be seen in the fractures or in the trenches is quite inadequate to demonstrate their presence in a more or less uniform and continuous layer. We shall continue to follow the Hess hypothesis, although the oceanic layer certainly results from more complex combinations than his 70% serpentinization; only drilling can settle the matter.

4. Transverse Fractures and the Seismic Anisotropy Due to Sea Floor Spreading

Seismic studies have shown that the transverse fractures correspond to a radical change in thickness of the oceanic layer (Figure 69), which would exclude the hypothesis of a vertical throw, if this were necessary. But the movement they reveal extends much deeper, for a very odd anisotropy in the propagation of seismic waves within the oceanic upper mantle has been observed.

Seismic anisotropy is generally a local phenomenon, more often resulting from macroscopic schistosity than from a preferential orientation of the grains. Hess (1964) notes however that olivine is readily oriented by flow or deformation (banded dunites, in which the (010) plane of the olivine is the plane of best cleavage and most probable slip). Now, the P-wave velocity changes from 7.7 km/s for the (010) plane to 9.9 for the (100) plane.

Hess takes from Raitt the velocities under the Moho corresponding to 14 refraction profiles taken in the vicinity of America, within a few degrees of the east-west Mendocino fracture. Figure 70 shows that the velocity has a sharp maximum when the propagation is parallel to the fracture, and a minimum when it is parallel to the magnetic anomalies. The figure is augmented by utilizing seven profiles of Shor, taken near Hawaii, immediately north of the Molokai fracture.

Velocities under 8.0 km/s have been left out in Figure 70; however an anisotropy in the same sense occurs in the anomalous mantle near the Gorda ridge.

Moreover, examination of mylonitized peridotites of the St. Paul Rocks has shown that most of the grains were oriented in a way to make the direction of minimum index, and hence of maximum seismic velocity, perpendicular to the plane along which the rock had been deformed.

Finally Sugimura and Uyeda (1966) credit the same anisotropy with a direct influence on the first motions of deep-focus earthquakes.

These observations show clearly that sea-floor spreading is not confined to the basement, or even to the oceanic layer, but involves the upper mantle.

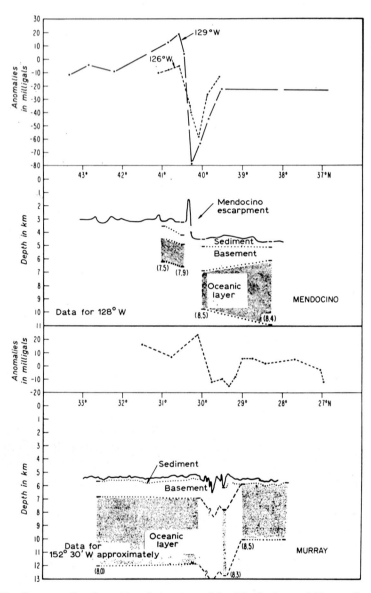

Fig. 69.   Gravity anomalies (free air) and structure of the Mendocino and Murray fracture zones.
(Menard, 1964.)

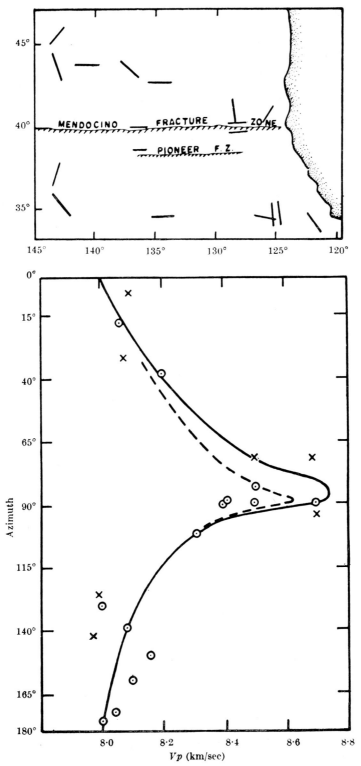

Fig. 70. *Above*, location of Raitt's refraction profiles. *Below*, *P*-wave velocity as a function of azimuth. Circles and the *dashed* curve concern Mendocino; *crosses*, Maui; the continuous curve, combined regions. (Hess, 1964.)

5. Gravimetric Study of Ridges and Grabens

Gravity measurements (Worzel, 1965; Harrison in Runcorn (ed.), 1967) provide information of a different type; unlike seismic profiles, they give no direct information on layered structures, but they are sensitive to horizontal variations, for example in passing from a basin to a ridge or trench.

The first gravity observations from a submarine, due to Vening Meinesz, date from 1923. Continuous profiles are now obtained conveniently from surface vessels, and will soon be available from airborne gravimeters. From 1935 it seemed clear that isostatic equilibrium prevailed for the oceans as a whole, and even between oceans and continents, the principal exception being the great negative anomalies along island arcs, discovered by Vening Meinesz himself. The progress of seismic studies has shown that this equilibrium corresponds fairly well on the whole to Airy's version of isostasy in which elevations are compensated for by the magmatic pressure on a depression of the Moho.

According to Harrison, the average characteristics of a normal oceanic region are as seen in Table II.

TABLE II

Layer	Thickness km	P-wave velocity km/s	Density g/cm³
Water	4.50	1.5	1.03
Sediments	0.45	2	1.93
Basement	1.75	5	2.55
Oceanic layer	4.70	6.7	2.95
Upper mantle		8.1	3.40

Densities, except for the first, are only known to within about 10%.

In order to study particular structures, a section obtained by seismic methods is taken as a starting point, and using empirical correlations (Birch; Nafe and Drake; Woollard, etc.) a density value is assigned corresponding to the P-wave velocity. Comparison between the gravity calculated from two-dimensional models and the observed results allows one to interpolate between seismic soundings, or at times, although this is more risky, to extrapolate in depth.

Vening Meinesz was already aware that the mid-Atlantic ridge, like the ocean basins, was roughly compensated isostatically. Consequently the 'free-air anomalies' which are computed from measurements at sea are small on the ridges (in contrast, the Walvis and Rio Grande rises have strong positive anomalies). These free-air anomalies depend directly on the undersea relief. To minimize this dependence and so make the density of deep structures accessible for study, geophysicists make use of another kind of anomaly, the 'Bouguer anomaly'. (This consists in assuming schematically that the rocks constituting the relief have a constant density of 2.6 or 2.7 g/cm³, for

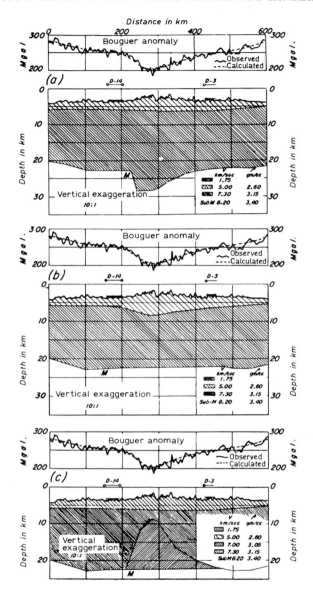

Fig. 71.   Three models of the crust under the mid-Atlantic ridge near 30°N which satisfy gravimetric data. (Talwani, Heezen, and Worzel; 1961, *Publ. Bur. Centr. Sismologique A*, **22.**)

example, and then calculating the effect of replacing all the water by rock of this density.) The result for a section of the mid-Atlantic ridge at 30°N appears in Figure 71. The Bouguer anomaly over it has a pronounced minimum, proving, as was to be expected, the existence of lighter masses under the ridge than under the lateral basins.

In a 1961 paper, Talwani, Heezen and Worzel had considered three hypotheses on the nature of these compensating masses: projection at the base of the crust, projection at the bottom of the basement, slight intrusion under the lateral basins. Their models assumed anomalous mantle under all the ridge. It is known now that the flanks are underlain by normal mantle; this spoils the agreement obtained. To restore it, Talwani *et al.* (1965) suppose (Figure 72) that the anomalous mantle indeed exists under the flanks, but in the form of a wedge concealed in the normal mantle; like all low-velocity layers, this wedge could not be detected by seismic refraction.

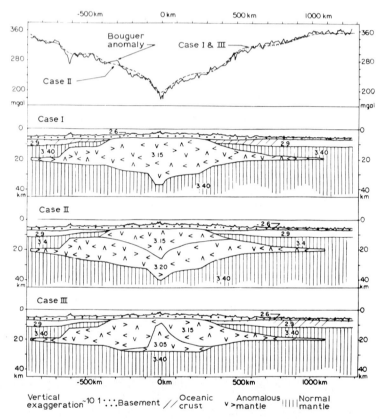

Fig. 72.   The models of Figure 71 modified to allow for return of normal mantle under the ridge flanks. (Talwani *et al.*, 1965.)

Data for the East Pacific rise, though less complete, are not opposed to the presence of similar wedges, probably restricted to the oceanic side in sections where the ridge follows the edge of the American continent.

The parts of ridges strongly perturbed by transverse fractures should in principle be excluded from the considerations above. Here there can no longer be any question of isostatic equilibrium, as the Meteor expedition of 1965 demonstrated for the Romanche trench. However approximate isostatic equilibrium may obtain in the neighborhood of the great fossil fractures of the East Pacific, if judged by the case of Mendocino, studied by Dehlinger *et al.* (1967), over 450 km of its length starting from the Gorda ridge. Raitt had already shown, in 1963, the presence of anomalous mantle in this region. The fracture divides a plateau at 3 km depth on the north, from a gentle slope dipping down to 4 to 5 km to the south. The free-air anomalies (a few tens of milligals) correspond roughly to the edge effect of an isostatically compensated plate, and not to that of an escarpment in a homogeneous medium. A complete calculation enables one to assign a high velocity to the mantle in the southern region, while at the same time showing it to be very abnormal to the north under the Gorda ridge; the density contrast seems to drop from 6% to 1% on reaching the western limit of the region under study, where the escarpment itself tends to disappear. It would be interesting to see if this gradual fall in the density contrast could be reinterpreted here also as a tapering off of a wedge of anomalous mantle.

Let us now leave the ridges to turn our attention to the continental rifts which cut across high plateaus, such as those of the great African lakes or that of Lake Baikal, in short to structures where the first stage in the formation of a ridge is seen. These rifts show anomalies without wings, which may be interpreted schematically by the subsidence of a prism between two normal faults (or two systems of faults). Agreement on this view was reached after much discussion. The prism is subsequently supported isostatically by the magmatic pressure on its lower edge (Figure 73).

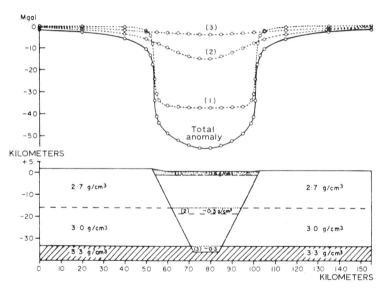

Fig. 73.   Bouguer anomalies of a graben on the hypothesis of a subsided prism isostatically supported. (1), (2), (3) respective effects of the upper, middle and lower layers, with density contrasts of −0.6; −0.3; −0.3. (Girdler, 1964.)

The second stage of the extension, after the grabens, corresponds to a rupture of the crust at an oblique angle to the structure, according to Girdler (1968) for example. The Red Sea, the Gulf of Aden or the Gulf of California are examples.

In the Red Sea, the existence of a strong positive Bouguer anomaly over the axial

trench (Figure 20), as well as the presence of the magnetic anomalies of course, gives
evidence of the rising of dense basalts up to the hydrostatic level, thus maintaining
isostatic equilibrium. Allan and Pisani (1966) have completed these older results of
Girdler. The Bouguer anomaly coincides with the axial valley. Where the latter is
most fully developed, the anomaly rises to 150 mgal. The magnetic anomalies disappear around 23° N, but the Bouguer anomaly remains positive nearly all the way
to the Gulf of Aqaba, 2000 m deep, where it falls to − 100 mgal.

The Gulf of Aden (studied by Laughton) has a fairly similar though not identical
structure (Girdler, 1968); in the Red Sea the crust on both sides of the axial intrusions
is continental: the Gulf of Aden has an oceanic crust throughout. Girdler considers
it possible that the separation of the Gulf of Aden occurred at three times the rate of
the Red Sea, which would entail a distortion of Africa if Arabia is considered to be
a rigid plate (Figure 74). Geologists however regard such a simplification with mis-
trust. Dubertret in particular insisted in 1967 on the great age and complexity of the
related structures observed in the Dead Sea rift.

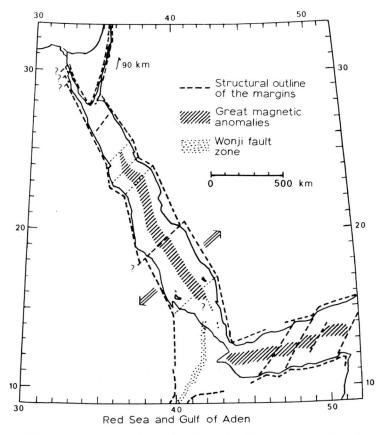

Fig. 74.   Possible movements of Arabia and Africa according to the transform faults and the
magnetic anomalies. The Wonji fault in Ethiopia after Mohr. (Girdler, 1968.)

Harrison and Mathur in 1964 found a Bouguer anomaly of 100 mgal over the axis of the Gulf of California, which correlated with a deep valley, and was interpreted as a strip of dense crust.

To summarize, consider what are thought to be the three successive extension structures; a minimum in the Bouguer anomalies corresponds to the slumping of continental grabens over widths of some tens of kilometers; next, at the Red Sea stage, an axial maximum appears due to the intrusions following the splitting of the crust; finally, in the ridge stage, again a low in the Bouguer anomaly, this time very broad, attributable to the anomalous mantle, but the gravity high over the axial valley has paradoxically disappeared despite the continuance of intrusive phenomena.

6. Continental Margins. Complete Study of Trenches

Knowledge of gravity anomalies (which exist on both sides of the coasts even if isostatic equilibrium is realized) has added to what we know of continental margins. Figure 75 shows that a satisfactory agreement can be reached in this way with the seismic soundings where they are available. In all the cases studied, the rise of the Moho on leaving the continent (or major oceanic islands) is at first very rapid, with changes of around 20 km for 100 km distance perpendicular to the coast, the slope being steepest around the 1000 fathom isobath. The Moho then rises more and more slowly towards its oceanic depth.

Advantage will now be taken of gravimetric data in an attempt to clear up the question of trenches introduced in Chapter I. Heezen (in Runcorn (ed.), 1967) divides them into three categories: the trenches of the island arcs bounding an epi-continental sea; the straighter trenches bordering a continent or its equivalent (Peru, Central America, Tonga); lastly the hair-pin shaped arcs of the Antilles and the South Sandwich Islands. These last are not however the only ones to be extended by regions of horizontal displacement or transform faults arising from the ends of the arc; it has been seen in Chapter V how in the Aleutians, for example, overthrusting gradually gave way to horizontal displacements as the western end was approached.

It has already been mentioned that trenches are associated with a deep gravity low (100 to 200 mgals) confined to a narrow strip (100 to 200 km wide). Vening Meinesz took the cause to be the formation of a fold in the light crust which was being driven into a dense magma by a lateral compression. This appeared to apply not only in cases where the strip coincides with the trench, as to the north of the Antilles, but also in the case of double arcs as in Indonesia, where the strip coincides with the more or less emergent exterior sedimentary arc which is supposed to be pinched in the root-fold. Other circumstances may occur, with the gravity measurements showing a secondary minimum as at Guam, or a minimum offset from the trench in the direction of the arc as in the Aleutians. The theory accommodated itself to them.

The combination of seismic refraction with gravimetry allows us to describe all these cases without appealing to the apparently successful idea of a root-fold (the first blow to the Vening Meinesz theory was struck by Ewing and Worzel in 1954

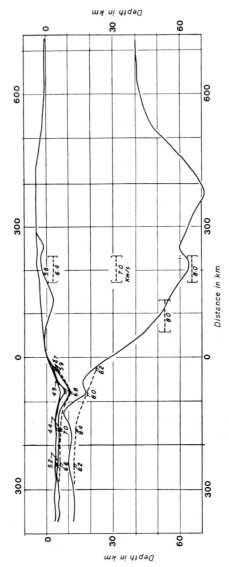

Fig. 75. Crustal section perpendicular to the continental margin and trench near Antofagasta (Chile) from the gravity data (Wuenschel, unpublished thesis) compared with the more recent seismic results of Fisher and Raitt. (Worzel, 1965.)

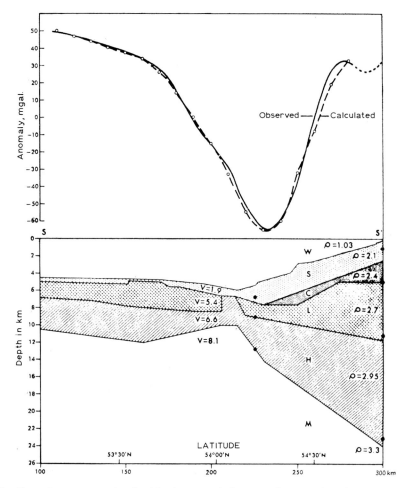

Fig. 76. Crustal structure under the Aleutians trench from gravimetry, seismology and magnetism. *Above*, free air anomalies (Peter *et al.*, 1965.)

in their study of the Puerto Rico trench). As an example of the structures thus obtained, Figure 76 shows that of the Aleutians trench after Peter *et al.* (1965). The gravity minimum would coincide with the deepest part of the trench if the mass of sediments accumulated on its northern slope were taken away.

Arguing from the gravimetry, Worzel (1965) interprets all trenches as slump grabens, which consequently bear witness to a slight extension (a few kilometers), and not to a compression of the crust. In order to explain the number of continental margins without trenches, he supposes the latter to be posterior to the development of continents and chains of islands.

This idea of extension, which agrees with the morphological observations of Menard (Chapter I), runs into an objection: in opposition to what we have seen for the continental rifts, the strips of Vening Meinesz are bordered by wider positive anomalies of almost the same intensity, which his theory endeavored to explain. Today a general explanation of these is not known; each individual case must be attributed to nearby relief, or, on the contrary, to the compensation of deep trenches.

A complete study of the Peru-Chile trench was published by Hayes in 1966. He suggests that it corresponds to a steeply-dipping normal fault starting from the base of the precontinental rise, while in the open sea a flexure with thinning of the crust should be found. Hayes concludes from this that compression is absent. Besides, the east flank is less rough than the west, and seems to lack the sediment traps which, according to Menard, restrict the filling-in of the trenches. The east flank is likewise much more magnetically quiet than the west flank. Magnetizable materials, or at least their relief, must therefore be different on the two sides. Extension and dissymmetry are equally in accord with Figure 76, where a positive magnetic anomaly near the trench axis is attributed to a fissure of the crust 20 km wide and 300 km long filled with the basic matter of the basement. Finally let us mention the recent study of the trenches west of Luzon by Ludwig et al. (1967) and Hayes and Ludwig (1967); it concerns a double trench bordering the *inside* of the Philippines arc, on the side of the epicontinental sea, thus recalling the Bismarck, Solomon or New Hebrides arcs (other such arcs undoubtedly remain to be discovered). Here again the east flank is more inclined than the west, at least on the average, for the latter has some steep slopes suggesting normal faults.

The law of dissymmetry which thus seems to emerge unfortunately has exceptions; there are trenches with regular flanks which are almost symmetric (Kermadec).

Returning now to sea floor spreading, the dissymmetry of trenches corresponds nicely to their role as sinks for the oceanic crust at the end of its course, but one would expect to see it accompanied by a compression of the kind held responsible for the folding of the nearby sedimentary arcs. Now, folding is not observed in the trenches themselves, and we have just seen that the morphology suggests the presence of an extension instead. To escape from these difficulties, Elsasser (1968) has suggested that the forced descent of the lithosphere under the trench would rapidly initiate a transformation of the matter to a denser form, which would speed up the process of descent beyond a certain level, thus exerting traction on the lithosphere and drawing out the flanks of the trench. Details of this ingenious suggestion, particularly its connection with the loss of water experienced by the matter, remain to be filled in. Oxburgh and Turcotte (1968) have tried to prove that friction during the descent would suffice to melt the basalt and to produce andesitic volcanism.

If the crust is engulfed in the bottom of trenches, even without compression, at least compacted and deformed sediments should be found there, if not folded ones. Reflection and refraction profiles have been made over some trenches, including sections thought to show strike-slip movement. Very generally (Pitman et al., 1968; Scholl and Von Huene, 1968; Hamilton and Menard, 1968) one finds a cover of pelagic sediments conforming to the basement, or of horizontally stratified turbidites, coming to an end over a sharp drop of the basement on the continental side; in short, structures corresponding to a prolonged calm rather than to a period of tectonic activity.

Magnetism provides a final series of facts for consideration. The field over the trenches is quiet. Although the number of linear magnetic anomalies located by now has increased considerably, none of them appear to penetrate the great trenches, even in a weakened form due to the engulfment; Figure 23 of Chapter II, completed near Hokkaido by Uyeda and Vacquier (Japanese National Report, 1967), and the fact that the final anomalies of the Lamont scale have been picked up near the Aleutians (Hayes *et al.*, communication at Zurich, 1967) merely allow one to suppose a prolongation of the known anomalies up to the immediate vicinity of the Northwest Pacific trenches. One is thus led to ask if the engulfment of the lithosphere did not take place *in front of the trench*, with the latter being the result of a secondary subsidence without deformation of the sediments. It would be idle to minimize the difficulties of this hypothesis (1).

(1) From Le Pichon:

The real problem (of trenches) is the remarkable absence of land-derived sediments in a trench near a mountain where erosion is very great. In regions where the trenches are filled, it can be shown that the rate of compression is lower and is thus entirely absorbed in the nearby zone of foldings. It must be remembered also that the Andes are in a strong rising phase. Finally, it must be noted that a mass of semi-consolidated sediments (velocity 2.5 to 3 km/s) is present on the internal edge of the trenches, where recent reflection profiles (Houtz, in press) show extremely close folds.

CHAPTER VII

HEAT FLOW THROUGH THE OCEAN FLOOR

1. World Heat Flow Distribution

The temperature of the Earth's surface and its variations with time depend on solar heating. However below some tens of meters of depth all these variations have disappeared except for effects of very long period, such as the after effects of the Quaternary glaciations. The temperature, constant in time, increases with depth. The temperature gradient between two points is obtained by dividing their difference in temperature in °C. by their difference in depth. It is of the order of some tens of degrees per kilometer, greatly variable with locality and depth.

The fall in temperature from the interior to the exterior of the globe corresponds to a heat flow in the same direction. This vertical flow by conduction is the product of the gradient and the thermal conductivity of the medium at the point under consideration. Heat flow is much more uniform than the temperature gradient, which varies in accordance with the differing thermal conductivities of materials. The flux ought to be expressed in units of power per unit area, but the custom of expressing quantity of heat in calories has persisted, so that heat flow is always given in microcalories per square centimeter of horizontal surface per second ($\mu cal/cm^2$ s).

Exact determinations of heat flow began on land, at Jeffreys' instigation, about 1939, and at sea towards 1952 at Bullard's. Results have been collected by Lee and Uyeda (1965) from whom we take what follows. By the end of 1964 about 2000 values were available, almost all oceanic and more recent than 1960: a good half were satisfactory with instrumental errors under 10%, but with a far from uniform distribution. In well-surveyed zones, regional variations over 0.2 $\mu cal/cm^2$ s are considered significant.

For the Earth as a whole the most frequent value (mode) is 1.1, and the arithmetic mean 1.58 ± 1.14 for 1150 measures, or 1.43 ± 0.75 if 389 mean values for $5° \times 5°$ squares are used. According to Lee, 1.5 may be taken as the general mean with 95% probability.

The differences between continents and oceans do not appear to be statistically significant. This fact was hailed as a great discovery when the first oceanic heat flow determinations were made. Goguel in 1957, then Crain in 1968 pointed out that the continental average needed to be raised appreciably to allow for the superficial cooling of large areas during the last glaciation (the same effect would go some way to explain the low values of flux in the shields, where drilling is usually shallow because of the hardness of the rocks). Let us assume however that the equality is indeed real.

It is then surprising, for the continental crust, containing granite, a highly radioactive rock relatively speaking, ought to have a higher flow.

The reason generally given for the equality in mean heat flow is that radioactive atoms, not readily incorporated in silicate lattices, are squeezed out upwards during crystallizations, so that the radioactivity of continental granite derives from the mantle and is compensated for by the mantle deficit. We shall need to re-examine this question from the standpoint of mobile plates.

Let us go into some details. Table III includes the major geological divisions.

TABLE III

Region	Number of values	Mode	Mean	Standard deviation
1. Precambrian shields	26	0.9	0.92	0.17
2. Non-orogenic since the Cambrian	23	1.3	1.54	0.38
3. Orogenic since the Cambrian (excluding Cenozoic volcanic regions)	68	1.1	1.48	0.56
4. Cenozoic volcanic regions (excluding thermal regions)	11	2.1	2.16	0.46
5. Oceanic basins	273	1.1	1.28	0.53
6. Ridges	338	1.1	1.82	1.56
7. Trenches	21	1.1	0.99	0.61
8. Other oceanic regions	281	1.1	1.71	1.05

Heat flow values for the Precambrian shields cluster around 0.9. Fluxes are higher in the other regions, e.g. those with post-Cambrian orogeny; further, their distribution is irregular and unsymmetric, as is shown by the large dispersion, and by the difference between mode and mean. Similarly the heat flows for oceanic basins are fairly symmetrically grouped about a low mean, while the asymmetry becomes large for the ridges, where the mode is unchanged despite some very high fluxes found there.

To be exact, flows several orders of magnitude higher than those of the ridges are met with in thermal regions, but these, of limited extent, are excluded from the continental statistics where only values under 3 μcal/cm^2 s, for example, are retained.

The high mean for group 8 in the table comes in large part from the very high values from epicontinental seas within island arcs. Figure 77 gives an example for the Sea of Japan and Sea of Okhotsk. An analogous situation, with measured fluxes up to 5.6 (Sclater and Menard, 1967), is found on the plateau of the Fijis, an isolated block with depth and thickness midway between oceanic and continental, surrounded by trenches where heat flow is low. It is found also within continental arcs, to judge by the Carpathian basin, where various authors, notably Boldizsar for Hungary, have found values between 1.9 and 3.3, while the flow is normal or low over the flysch of the outer zone (Boldizsar, 1968).

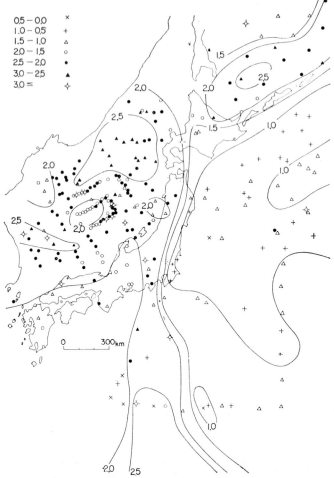

Fig. 77.   Heat flow in μcal/cm²s in the neighborhood of Japan (Uyeda and Vacquier: 1967, Japanese
National Report on the Upper Mantle.)

2. Heat Flow on the Ridges

Let us now return to our main interest, the ridges. The foregoing table shows that
the mean flow there is high. The flux is particularly high in the Gulf of Aden (3.9
μcal/cm² s), the Red Sea (3.4), and the Gulf of California (3.4), which should be
compared to a crestal zone rather than to a ridge as a whole. Less extreme heat flow
is found in continental grabens (for example up to 2.5 in the Baikal graben).

The distribution of values for a typical ridge, actually the East Pacific rise, seems
to be explained by a figure of Menard's (Figure 78) dating from 1960 which brings
together, in a presumably composite section, heat flow observations by Herzen,
Maxwell or Revelle, and a crustal section derived from seismic measurements of
Raitt or Shor. This diagram has often been reproduced, and clearly indicates a

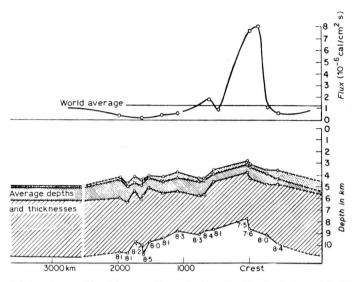

Fig. 78. Crustal structure and heat flow profile on the East Pacific rise (schematic). (Menard, 1960.)

correspondence between the ridge crest, the presence of anomalous mantle and the extraordinary heat flow values. It has since become evident that the true situation is much more complex. Figure 79 shows that in the region of the crest both high and low values are found, the very high values showing a tendency to cluster on both sides of the crest to a distance of about 150 km. Analogous circumstances appear on the Atlantic crest (Figure 80), but the group of high values extends only about

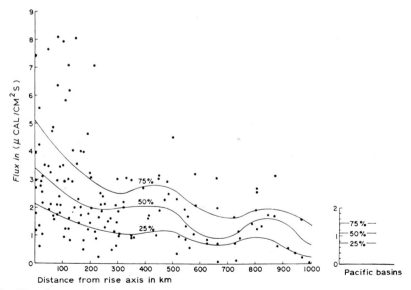

Fig. 79. Heat flow as a function of distance from the crest of the East Pacific rise (50°S to 20°N). Curves enclosing 75%, 50% and 25% of the values, for the rise, and for the Pacific basins. (Lee and Uyeda, 1965.)

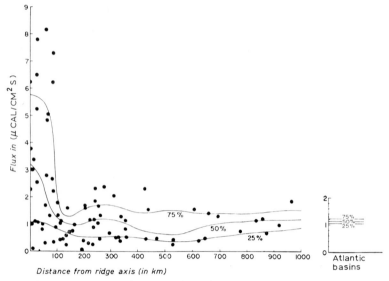

Fig. 80. Heat flow as a function of distance from the crest of the mid-Atlantic ridge. Curves enclosing 75 %, 50 % and 25 % of the values, for the ridge and the Atlantic basins. (Lee and Uyeda, 1965.)

50 km on either side of the crest, which is close to the possible error in position with respect to the axis (defined by its magnetic anomaly); at the same time, the average returns much more quickly to normal (without experiencing a marked depression on the flanks, as was once thought). The dispersion of the individual values remains much higher, however, than for the oceanic basins.

The presence side by side of high and low values is a general result for all the ocean ridges, including gulfs. The Indian Ocean furnishes diagrams similar to the preceding, but it has received particular study, most recently by Langseth and Taylor (1967). Their measurements of heat flow along the Indian ridge led them to reinterpret the structure in the rather speculative manner shown in Figure 81: high flux values are to be taken as indicating the vicinity of a segment of ridge, low values a transform fault. Preceding authors had found little correlation between flux and visible structures.

Finally, it is pointed out that values of heat flow may differ greatly on the two sides of the great Pacific fractures, which is hardly surprising, since the crust is known to have very different thicknesses, to say nothing of the upper mantle. Dehlinger *et al.* (1967) compare their gravimetric results (Chapter VI) to the flux observed by von Herzen in 1964 on either side of Mendocino (Figure 82). The mean heat flow is high over the anomalous mantle connected with the Gorda ridge, but the temperature excess deducible from this (on the order of 300° at 9 km depth) is insufficient to explain the density contrasts. These last are attributed by the authors to chemical transformations (serpentinization?) or to injection of dykes.

The variability of heat flow on the ridges has long been blamed on the errors of measurement peculiar to regions of highly irregular relief. In sea water the possibility of exchanges by *convection* assures a good uniformity of temperature at the irregular

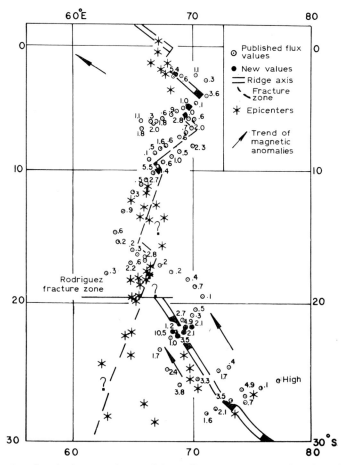

Fig. 81.   Heat flow in the central part of the Indian Ocean. (Langseth and Taylor, 1967.)

surface of the bottom. The same probably holds true, to a lesser extent, in unconsolidated sediments, or in fracture zones, as for example in the median valley of the
ridges. But generally heat flow may be thought of in terms of *conduction*, the flow
lines being perpendicular to the isothermal surfaces.

In the absence of sediments the flow lines would spread out on the humps, giving
lower heat flow there than in the hollows where they converge (Figure 83a). The
presence of sediments, whose conductivity is barely one-third that of solid rocks and
which are more abundant in the hollows, has an opposite effect in repelling the flow
lines (Figure 83b). It is this effect which perhaps explains some of the low values
found in trenches and in the median valleys of ridges, the more so since the tempera-
ture-measuring apparatus penetrates only in the sediments. It is generally difficult
however to sort out the opposing effects, although this has been attempted (Birch,
1967; Lachenbruch, communication at Zurich, 1967). An additional difficulty is the
existence of undulations of 100 m amplitude with 30° slopes, like those found by

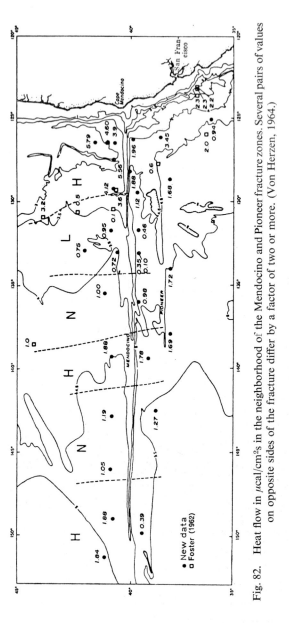

Fig. 82.   Heat flow in μcal/cm²s in the neighborhood of the Mendocino and Pioneer fracture zones. Several pairs of values on opposite sides of the fracture differ by a factor of two or more. (Von Herzen, 1964.)

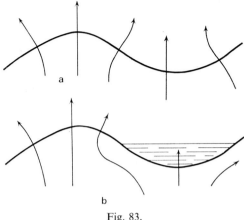

Fig. 83.

Loughridge in 1966 by means of a narrow-beamed submerged sounder, which may have an important role. Finally, differences of the order of 25%, from the preceding causes acting together, must be common, and differences of 75% may also occur in a small percentage of cases.

One may attempt to get rid of these differential effects by taking the mean of a certain number of observations, more numerous where the relief is more irregular. These means will still be too low (by 20% at most) for a new reason, which is that it is always the vertical gradient which is measured, not the gradient perpendicular to the bottom.

Even if the variability of heat flow values could be corrected exactly for the effects of relief, it would remain large, and would certainly not correspond to a smooth decrease in flow with distance from the ridges. Thus Sclater and Vacquier (1967), in the region between the East Pacific rise and the Gulf of California, and in the region between 10°N and 10°S including the Cocos rise, the Galapagos and the abyssal plain west of Nicaragua, find areas where the mean flow exceeds 3.5 over areas as large as 5° × 5° in latitude and longitude, surrounded by areas of lower flow. The opposite case of areas of systematically low heat flow had been observed earlier.

Langseth *et al.* (1966) have sought to compare the East Pacific and Atlantic ridges from the standpoint of the heat emitted by each, taking into account all trustworthy measurements made in two areas of similar extent (327 measures in the Atlantic and the Caribbean, 284 in the equatorial region of the East Pacific between 150°W and 70°W). The respective mean fluxes are 1.34±0.89 and 1.85±1.60, but more significant is the excess of the flow over its value in the surrounding basins. The quantity of heat furnished by the Atlantic ridge, reckoned as the total excess emitted on a 1400 km wide strip symmetrical about the crest, comes to 40 cal/cm s of ridge for the North Atlantic, 60 for the equatorial Atlantic, 0 for the South Atlantic. In the equatorial Pacific, account must be taken not only of the ridge but also of the Galapagos branch (Chapter IV), which is marked by particularly high values. If the total excess of heat is considered to be emitted over a section 3000 km wide, suitably defined to avoid a distinction between the two structures, 225 cal/cm s is found. There is thus a consider-able difference between the Pacific and the Atlantic ridge, supporting the idea that the latter represents a later stage of evolution; but this idea is far from proven.

Le Pichon and Langseth (1969) have completely reviewed the question of heat flow on the ridges, in particular the interpretation of the data. They assume that the conductive flux out of the mantle is defined by the mean of observations over an area whose dimensions are large in relation to the thickness of the crust. The number of values required for the mean varies with the region; on the ridges it is too large for the existing density of measurements. However it is possible to try to make the measurements taken at different points in different ridges more comparable by reducing the distance from the axis of these points to what it would be at a spreading rate of 1 cm/y for example. In order to leave room for a check, Le Pichon and Lang-seth make separate reductions for the East Pacific where rates go from 3 to 6 cm/y, and for the Atlantic-Indian ridge where they go from 1.5 to 2.2 (the authors further treat the segments of ridge between two successive fractures separately, in each case). To avoid the supposed discontinuities in rate prior to the Miocene, the reduction ratio is calculated from anomaly 5, and lastly the reduced distance D is assigned a coefficient such that $D=1$ for this anomaly.

In this way it is again found in general terms that the maximum heat flow does not occur right over the axis, but at some distance on either side; the flow then falls off towards the flanks; the fluctuations suggested in Figure 79 are not significant. Table IV gives the mean fluxes in various strips:

TABLE IV

D	East Pacific	D	Atlantic-Indian ridge
$0<D\leqslant0.54$	3.31 ± 1.94	$0<D\leqslant0.46$	2.72 ± 2.33
$0.54<D\leqslant1.8$	2.00 ± 1.29	$0.46<D\leqslant1.4$	1.45 ± 0.94
$1.8<D<3.0$	1.47 ± 1.03	$1.4<D<3.1$	1.10 ± 1.14

Two-thirds of the heat flow excess (over the adjacent basins) comes from a narrow axial strip younger than 6 m.y.

3. Interpretation of the Results

Let us now look for an interpretation of the variations in heat flow. One widely-used method is the calculation from all available data, despite the uneven global distribution, of the best fitting spherical harmonic series.

This series, up to a certain order, is then compared with a similar series calculated from the gravity anomalies, or equivalently, from the height of the geoid (surface of zero level) over the best-fitting ellipsoid. A weak negative correlation is found, which may be explained by the relief of the regions of recent tectonic activity, or, as many prefer to see it, by the effect of currents of convection. As Runcorn has pointed out, no very definite conclusions can be obtained on this view, because the heat carried by the currents could take 100 or 200 m.y. to diffuse; during this time the continents would have drifted considerably, so that the present situation would be the integrated result of the totality of past situations.

A purely convective model would also require the heat to rise under the ridge crest, which is very narrow and offset by transverse fractures. We have already shown in Chapter V, from a kinematic viewpoint, that this model may with advantage be replaced by one in which the convective zone is covered over by rigid plates whose separation under the ridge axis gives access to intrusions. The first model of this kind is due to Langseth *et al.* (1966). McKenzie (1967b) showed that a simpler model allowed a fairly good interpretation of the ridge observations. He treats the oceanic lithosphere as a non-radioactive plate 50 km thick with its surface at 0 °C and its bottom at 550 °C, in unperturbed regions. These numbers may be disputed; the thickness seems low, although McKenzie points out that it is close to the accepted depth for volcanic magmas, and to the depth of the top of the low-velocity layer in the Pacific, estimated at 55 km by Ewing, Brune and Kuo in 1962. The equilibrium gradient seems low too, at 11 °C/km, but let us accept it.

To reproduce the features of a ridge, the plate is split by a constantly enlarging straight fissure through which the 550 °C material is injected from the bottom to the surface. Each half-plate is supposed to have a quasi-stationary temperature regime with recovery to the equilibrium gradient taking place at an infinite distance from the fissure. The result is shown in Figure 84 for a spreading rate of 4 cm/y., applicable to the Pacific. It is seen how a vertical column, initially all at 550°, cools from the top as it moves away from the axis, and how the isotherms gradually fall to their limiting values.

Figure 85, derived using reasonable values for the density, specific heat and thermal conductivity, shows that the model well represents the average results of Lee and Uyeda for the East Pacific. Agreement would be equally good for the other ridges. The choice of a small plate thickness is essential, since the width of the heat flow anomaly varies with the square of the thickness. Of course, agreement quite close

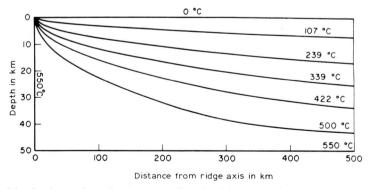

Fig. 84.   Isotherms in a plate moving off to the right at 4 cm/y. (McKenzie, 1967b.)

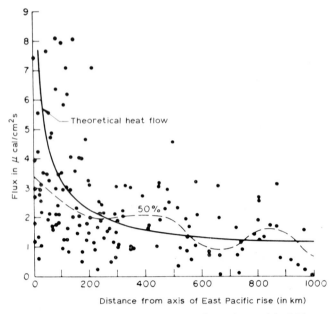

Fig. 85.   Heat flow in the Pacific. The theoretical flow from the model of Figure 84 is compared
with observation. (McKenzie, 1967b.)

to the axis cannot be expected, for the representation of the intrusions is too simple.
However the fit is generally good out to large distances, although the flux there depends
mainly on the hypothesis of constant temperature under the lithosphere. Hence
McKenzie concludes that the local conductivity must be sufficiently high to maintain
this constancy in the face of the deeper temperature differences necessary for any
convective theory. If a constant temperature over the surface of the low velocity
layer were general, it would have an important consequence: since this layer appears
to be deeper under the continents, the temperature gradient, and consequently the
conductive heat flow in the continental lithosphere would be lower. This effect might
compensate for the excess heat generated by the continental granite, and thus explain
the approximate equality of oceanic and continental heat flows.

In the more complex model of Langseth et al. (1966), the plate was fed by a progressive rising of material in the crestal zone. Lithospheric thickness (100 km) and
temperature at the bottom (1500 °C) seem normal*. Like McKenzie, the authors
suppose that the plate is not radioactive; they justify this at some length by showing
that any heat so produced would be negligible compared to that brought to the ocean
bottom by the overall convection; the plate thickness is deliberately taken large
enough so that the material which does not rise to it is not appreciably cooled. Later,
Le Pichon and Langseth (1969) showed that it would be interesting after all to assume
radioactivity in the plate. This would raise the heat flow curve at large axial distance

* The bottom is shown as impervious to heat in the figures of their paper, but this superfluous
condition was not used in the calculations.

without necessitating McKenzie's assumption of constant temperature for the base of the plate.

If the model of Langseth, Le Pichon, and Ewing were quasi-stationary, it would give too great a flow. The authors are thus led to assume that convection is intermittent, with the ascent of hot matter from the depths in 50 m.y. being replaced elsewhere by the descent of cold matter. Convection then ceases, and equilibrium is reestablished, also in 50 m.y. The results, for a spreading half rate of 1 cm/y., appear in Figure 86, which raises two important questions not considered by McKenzie: first we see that the ridge is uplifted, secondly that there is compensatory light matter under the crest, both effects produced by the thermal expansion resulting from a temperature excess in the axial region.

Le Pichon and Langseth (1969) have considered the quantitative conditions which a ridge model should satisfy in this respect. As Menard had already indicated, the

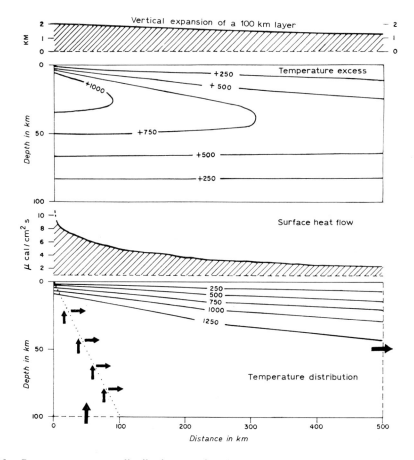

Fig. 86. *Bottom*, temperature distribution as a function of distance from the axis for a ridge with a spreading rate of 2 cm/y; *middle*, corresponding heat flow, *above*, excess of temperature over the equilibrium value in the absence of convective currents; *at top*, vertical expansion corresponding. (Langseth *et al.*, 1966.)

mean slope of a ridge (and also the roughness of its relief) varies inversely with the spreading rate, which is compatible with a simple plate model. Numerically, the difference in depth between a point on the ridge and the crest then depends only on the age of the ridge at this point. The authors find that the difference in depth between $D=1$ (anomaly 5) and $D=0$ (axis) is in fact constant, or nearly so, and equals 829 ± 112 m (Figure 87). Farther out, the reduced slope is less and its values more

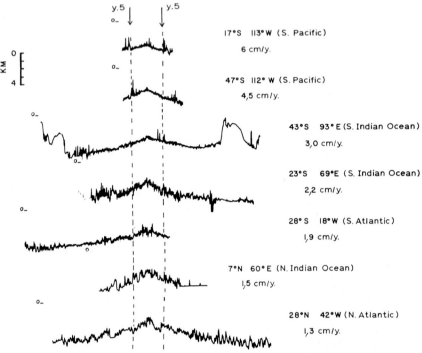

Fig. 87.   Bottom profiles in reduced distances for various oceans. (Langseth and Le Pichon, 1969.)

dispersed. In McKenzie's model the crestal temperature passes from an average of 275 °C to 550 °C. Assuming with Le Pichon and Langseth a coefficient of linear expansion of $3 \times 10^{-5}/°C$, the elevation of the ridge would be only 360 m, and the temperatures are too low to think in terms of a dilation due to partial melting. The model of Langseth, Le Pichon and Ewing, with twice the thickness, appears preferable here, and in addition, by the lateral penetration of high temperature materials, it furnishes the compensatory masses revealed by the gravity measures. On the other hand, the model slope is only half the measured slope.

Le Pichon and Langseth (1969) also recall the viewpoint of Orowan (1966), according to which ridges are produced by the rising of a hot mass in a plastic medium. On approaching the surface the mass splits and the two parts separate with a stagnant zone developing between them over the rising current. *Stagnant zone models* may be constituted by assuming that the system thus defined is covered by a rigid lithosphere.

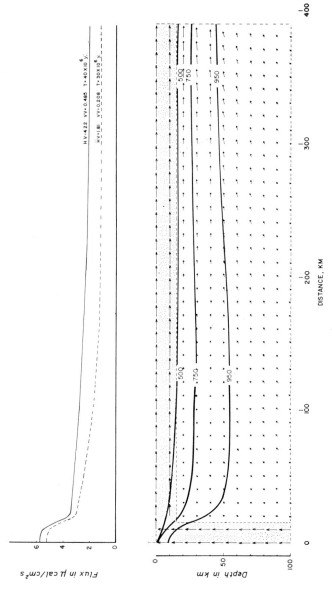

Fig. 88. Isotherms *(lower)* and heat flow *(upper)* in a stagnant zone model. *Full line* flux for the East Pacific rise; *dashed*, for the Atlantic-Indian Ocean ridge. (Langseth and Le Pichon, 1969.)

According to Elsasser (1968), the stagnant zone is wide if the depth of convection is great, but the ductility of serpentine allows the plate to slide. A supplementary local convection in the stagnant zone may also be assumed, with von Herzen and Uyeda.

Figure 88 corresponds to a schematic model of a stagnant zone: a plate only 15 km thick moves with a velocity of 4.22 cm/y. (East Pacific rate). Between 0 and 15 km from the axis, the vertical velocity is constant at all depths. Beyond this, up to 390 km from the axis (about $D=1$), the speed of ascent measured at 120 km depth is uniformly 0.485 cm/y. The horizontal movement speeds up a little with decreasing depth and increasing distance from the axis until it reaches the plate velocity. The temperature at 100 km is 1000 °C.

For a spreading rate of 1.81 cm/y. (Atlantic-Indian rate), the velocities should be reduced proportionately, but the dimensions of the stagnant zone would be the same.

The velocity being low at depth, the time needed to approach temperature equilibrium is much longer in these models than in simple plate models, reaching tens of millions of years. Figure 88 shows the surface flux reached in 40 m.y. for the East Pacific and in 30 m.y. for the Atlantic-Indian ridge. Agreement is satisfactory. A composite model of this sort, unlike simple models, permits a heat flow anomaly dependent on the spreading rate, but as it involves 5 parameters, its adaptability is not surprising.

Temperature distribution in this stagnant zone model still does not allow the relief to be explained by simple dilatation. The elevation would pass through a broad minimum towards 100 km. Besides, the relatively low temperatures near the axis would not explain the shallow depth of the observed compensation. The model would be improved by taking a thicker lithosphere, higher vertical velocity in the axial zone, and above all, by assuming partial melting of the materials. As Bott (1965) had clearly seen, it seems that no completely satisfying model for the elevation of ridges can be obtained without postulating phase changes in their axial zone, partial melting being the likeliest change.

Let us point out in conclusion that if the topography of the ridges is a phenomenon of thermal dilatation (complicated by phase changes and isostatic adjustments), it must show traces of the fluctuations in spreading rates (Schneider and Vogt, 1968). If the rate is uniform, a normal form of ridge results, corresponding to the progressive cooling of the lithosphere with distance from the axis. If spreading is halted, the ridge slumps; if it starts up again, projecting plateaus should be formed, as is actually observed in both sides of the axis in the crestal zone of the North Atlantic. The mountainous aspect of these plateaus would be due to the low new spreading rate (cf. Chapter VI). Thus the interruption in spreading, of Ewing and Ewing (1967), at the end of the Tertiary is again found, even if no more exact definition is furnished for it.

CHAPTER VIII

NATURE OF THE UPPER MANTLE AND THEORIES OF CONVECTION

1. Thermal Convection the Cause of Sea Floor Spreading

We have assumed that convection currents in the Earth's mantle are the motive force for the lithosphere. The geological literature has often invoked such currents, each time endowing them with the peculiar properties required for the problem at hand, without stating explicitly their general characteristics or their deeper causes. The aim of the present chapter is to show that provisionally acceptable theories may now be drawn from the very great variety of those put forward.

The introduction of a rigid plate avoids the necessity of attributing the median valleys or the transverse faults of ridges to local details of the convective system: the zig-zag pattern of crests and faults may be viewed as reproducing in detail the initial tear, as Wilson (1965) does; on the other hand, the general course of this rupture, and especially the location of the descending trenches, reflect the main trends of the underlying convection. However it is not easy to visualize the arrangement of the currents.

Hess (1965) adopts the 'tennis-ball' model (Figure 89), in which two convective cells are joined together by their descending walls, which correspond to the seams of the ball, while rising occurs along the median arcs of each of the two constituents. We have seen, indeed (Chapter IV), that the ridges form two continuous systems, from the mouth of the Lena to Bouvet Island, and from Alaska to the Gulf of Aden, if the south-west branch of the Indian ridge is assumed to be extinct. In each cell

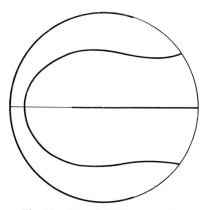

Fig. 89. 'Tennis ball' convection.

the spreading rate would vary simultaneously over the whole length of the ridge. The system of descending lines is supposed to be continuous and closed; this is not too evident, but we shall return to it (Chapter IX).

It will be noted that in the model the rising lines have free ends; but in at least one case, (the Gulf of Aden), ridges continue into bifurcated continental grabens. It is possible that the rising current weakens in approaching the free end, as occurs in experimental convective rolls: the floor spreading originating from the other arc in that case should push the nearest descending region towards this free end. It would be premature however, to apply such a suggestion to the South Atlantic, where the free end near Bouvet Island occurs. Besides, according to Hess, arcs with a free end should be unstable; this is how Le Pichon (1968) explains the movement of the South Atlantic ridge into a median position (Chapter V).

Putting topological difficulties aside for the present, let us consider the internal phenomena. If the convection involves the whole mantle (thickness 2900 km), even if the relations between the superficial and the deep currents are very complex, it seems that very considerable heterogeneities at the boundary of the core would be needed to give the surface distribution its observed geographical complexity. As such heterogeneities are rather unlikely, we shall favor theories that restrict convection to the upper mantle.

We shall assume this convection to be essentially of thermal origin. The energy liberated by the formation of the crust is small: Elsasser (1968) calculated that if one-half percent of the upper mantle (down to 800 km) comes to the surface in 1000 m.y. and if a density difference of 0.5 g/cm^3 is involved, the energy corresponding is 4×10^8 cal/cm^2, compared with 4×10^{10} deduced from the present heat flow. The effects would not be comparable unless the efficiency of the thermal engine fell to 1%; this, as Stacey points out, would barely furnish enough energy for earthquakes. Nevertheless differentiation phenomena may play a certain role.

We shall argue later as though hydrostatic equilibrium and a stationary thermal regime were not much perturbed by convection, but the changes in the arrangement of convective regions in the course of geological epochs must stem from non-linear and probably irreversible effects. It is in this perspective that Runcorn regards orogenic cycles as the effect of successive adjustments of the geometry of the convective cells to the thickness of the mantle, which would still be diminishing at present in consequence of the descent of iron into the core. This explanation no longer holds if convection is restricted to the upper mantle, but it could apparently be adapted to the superficial differentiations which cause the thickness and composition of the upper mantle to vary.

In a general way, it would be very useful to be able to connect present convective movements with the historical evolution of the Earth. This evolution is unfortunately very hypothetical, mainly as a result of our ignorance of the internal radioactivity. As a matter of fact, the heat given off by the granites could account for nearly all of the present continental heat flow. Radioactivity is definitely lower in basalts; the tholeiitic basalts, whose extent in ocean floors has been mentioned, are nearly three times less radioactive than the other basalts, and peridotites a hundred times less. To estimate the present importance of superficial heat sources, one should start from the seismological structure of the crust. But the first stage in a reconstruction of the past would be an extrapolation in depth, which is most hazardous. It will be seen in this chapter that the major pieces of information which can be used derive from indirect considerations on the state of the matter involved, particularly its proximity to the melting point.

Let us attempt nevertheless to sketch a possible evolution for the primitive Earth. For a long time the Earth was believed to have condensed in liquid form; today the assumption that it was formed by accretion of solid particles is preferred. However the gravitational energy liberated by impact and settling, added to the higher radioactive heating of that epoch, which would have difficulty in escaping from a poorly conducting Earth, may have melted it. We make the classic assumption, then, that it has been liquid, while referring the reader to a remarkable article by Lubimova (1967) for a contrary view.

Cooling at the surface, together with the sinking of heavy materials (essentially iron) to constitute the core, must have stirred the mass sufficiently to make its different parts interchangeable. In making comparisons between parts at different depths, no heat is supposed to be lost in bringing them to a common level; this amounts to saying that the temperature varied with depth according to the adiabatic law; the corresponding temperature gradient may be estimated at a few tenths of a degree per kilometer. As convection is a powerful means of heat transport, the temperature fell throughout the mantle, maintaining the above gradient up to the moment when the melting point was reached and solidification began. The slope of the curve of melting-point as a function of depth has been estimated also. It is of the order of 10 times the adiabatic slope; from this it may be deduced, with Adams and Williamson (Figure 90),

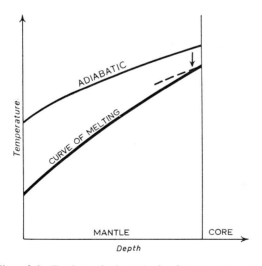

Fig. 90.   Cooling of the Earth on the hypothesis of passage through a liquid phase.

that solidification began at the base of the mantle* and was propagated toward the surface from that point. At the same time, and this is very important, the radioactive atoms, not readily incorporated in the crystal lattice of the silicates, were driven upwards without allowing time for the energy they liberate to play an important part in the phenomena as a whole (Jeffreys). At present the lower mantle, solid and non-radioactive, is supposed to be immobile, but the upper mantle would still be capable of a weak thermal convection (while still transmitting seismic waves of short period).

This description is merely indicative: the word convection is to be understood in a very broad sense, without implying that the corresponding currents are stationary. To define the words liquid and solid, melting and solidification for the complex mixtures probably constituting the mantle would be difficult. Even with pure constituents, as Goguel (1965) noted, it would still be necessary for the heat of crystallization of the last liquid portions to be dissipated by conduction, requiring a very long time; Goguel envisages the material as a fluid much thickened by crystals, solidified in places while still mobile in others.

Going no further with these historical considerations, we next seek to clarify our ideas on the present state of the upper mantle by outlining this extraordinarily involved subject as briefly as possible.

2. Properties of the Upper Mantle Given by Seismology

The most detailed data we possess about the interior of the Earth are on the velocities V and W of the longitudinal and transverse waves which travel through it. Assuming that the velocity increases with depth**, it was found from the first seismological observations that the increase was rapid, if rather irregular, down to about 900 km, and much less rapid farther on. This defines the depth of the upper mantle approximately, for our purposes.

In a homogeneous isotropic elastic medium, the respective velocities of the waves P and S depend on the bulk modulus k, the rigidity μ, and the density ρ according to the formulas $3\rho V^2 = 3k + 4\mu$, $\rho W^2 = \mu$. From a knowledge of V and W as functions of the depth, Bullen, and later others, have deduced the density, after recourse to various hypotheses. These usually include the two following: the medium is approximately in hydrostatic equilibrium; as long as its composition is unchanged, the variation of pressure as a function of density corresponds to the adiabatic bulk modulus which appears in the formulas above. Whatever the method used, the density variation is found to be analogous to that of the velocities: it increases by about a third down to 900 km, and another third by 2900 km. The elastic moduli also vary in the same fashion.

Gutenberg was the first to present serious arguments casting doubt on the continuous increase of seismic velocities with depth. But the existence of a layer in the upper

* The argument has been completed by Jacobs by bringing in the core. See for example Coulomb and Jobert (1963).
** Actually the velocity gradient must be compared with the ratio of the velocity to the radius, to allow for the spherical shape of the Earth; we pass over this detail.

mantle in which either P or S velocity might diminish was only well established by later studies of the dispersion of long period seismic waves (Love and Rayleigh waves) which are propagated horizontally, following the curvature of the earth. In what follows we shall speak, in any case, as is usual, not of a "layer with diminished velocity" but of a "low-velocity layer", that is to say, we shall consider it to be thick enough for the velocity in question to return to the value it had before diminishing.

In the last 15 y. important information has been gained from study of the vibrations of the earth as a whole set up by very large earthquakes. Their periods (which go up to 54 min for the lowest mode) may be obtained by harmonic analysis of the records of special seismographs. At present values are known for about a hundred of these periods, and they may be used to improve the velocities or densities obtained on the classic hypotheses. Figure 91 shows the 'S Model' derived by Verreault (1966)

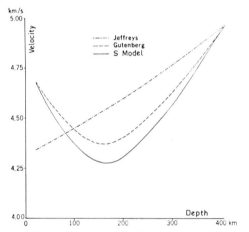

Fig. 91.   Velocity of transverse waves in the upper mantle. (Verreault, 1966.)

from torsional oscillations, compared with the curve for the velocity W of S-waves obtained by Jeffreys in 1939 (no low-velocity layer involved) and Gutenberg's 1959 relation. Verreault is led to give more importance to the low-velocity layer than Gutenberg. However a confirmation of this result would be desirable.

Figure 92 compares the density in the S model to that of two earlier models: Bullen A, obtained in 1953 from Jeffreys' velocities, which as we have just noted involve no low-velocity layer; and Birch, of 1961, which even shows a minimum in the density; the low-velocity layer would then be mechanically instable. This is not the case for Verreault's model. The presence of a minimum in the density requires a very rapid decrease in the rigidity μ, while it need only be assumed to increase less rapidly than ρ for W to diminish. As a less pronounced low-velocity layer is always found for V than for W, the conditions to be satisfied by the modulus k are even less restrictive than for μ. It is not necessary therefore, to suppose that the mechanical properties of the matter of the low-velocity layer are much different from those of its surroundings.

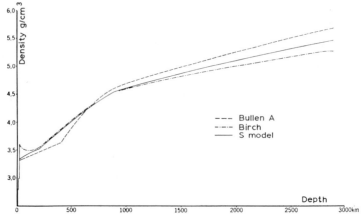

Fig. 92.   Density in the mantle. (Verreault, 1966.)

However they are generally assumed to resemble those of a viscous liquid in damping seismic waves and having low rigidity.

The boundaries of the low-velocity layer are still very poorly known. The models of Gutenberg and Verreault we have just seen; other estimates conclude in a more rapid recovery. In 1961 Miss Lehmann put the frontiers at 135 and 200 km, the S-wave velocity rising sharply at the lower depth (Lehmann, 1967). Mohammadioun (1966) finds the spectrum of P-waves to be deficient in short period components as a result of absorption when the seismic focus lies between 140 and 235 km deep or when a large part of their path is in this layer. The agreement on depths seems excellent. On the other hand, many authors from a study of the imperfections of elasticity of the mantle, or rather its quality factor Q, have the layer going down to about 400 km.

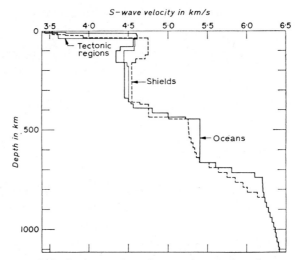

Fig. 93. S-wave velocity in the mantle under oceans, tectonic regions or shield areas.
(Anderson, 1967.)

Actually the thickness of the layer must vary with the region. Anderson and Toksöz (Anderson, 1967) have obtained the preliminary results indicated in Figure 93 by comparing, from the same earthquake, spectra of Love waves which have made different numbers of turns around the Earth before arriving at the same place. The differences between oceans and shields appear to be significant down to 400 km, perhaps to 800, while regions with mountain ranges would have intermediate values. If all the layers above the oceanic low-velocity layer are taken to constitute what we have called the lithosphere, it should thus be assigned a thickness of about 100 km, as we have done.

Seismology has provided other results. Birch has shown that seismic data would be hard to reconcile with the theory of finite deformations in isotropic bodies if an invariable composition was assumed for the mantle between 200 and 900 km (as for the lower mantle, it may be chemically homogeneous). Then too there have been repeated announcements of the discovery of discontinuities or zones of rapid variations of seismic velocities in the upper mantle. The installation of extensive groups of seismographs interconnected as *arrays*, has recently led to the proposing of new subdivisions (Anderson, 1967) at various depths depending on the region. The results do not seem to be definitive, and the existence of clear-cut discontinuities may be doubted. We shall have to ask ourselves however whether heterogeneities are compatible with convective mixing.

Lastly it may be recalled that the statistics, so often invoked, on the number of earthquakes as a function of depth, are somewhat inconclusive. Few are found between about 300 and 350 km, and it is difficult to attribute this to the low-velocity layer, especially when according to Knopoff (1967) the minimum goes down to 400 or 450 km if the total energy of the quakes is considered instead of their number. None are found below 740 km.

3. Temperature and Composition of the Upper Mantle

If the temperature gradient of $20°$ to $30°$/km observed in drillholes continued at depth, matter would certainly be molten at 100 km. The gradient accordingly must begin to fall off beyond a certain point. But the exact circumstances are very poorly known. Aside from an attempt by Verreault (1966) to deduce the temperature from his density model, the only relatively direct way would be to compare laboratory studies of the *electrical conductivity* of the silicates assumed to be present in the mantle with the conductivity calculated for different depths from the surface variations of the Earth's magnetic field (Rikitake, 1966). Unfortunately neither of the terms to be compared is well determined at present. Using the magneto-telluric method various authors including Grenet in France find an initial increase of conductivity (in order of magnitude it would change for example from 10^{-14} emu to 10^{-13} or 10^{-12}) at around 70 or 80 km; then, after having probably decreased, by 600 or 700 km it would rise again, rapidly reaching 10^{-11} emu (Price, 1967). On the laboratory side the experiments have been carried out on only a few minerals and at much too low pressures; ionic conductivity is important at first, then the mantle must begin to resemble an electronic semiconductor. Thus of the two conductivity modes, one would be predominant in one domain of observed increase, the other in the second, but their respective importance is poorly known (Price, 1967; Lubimova, 1967).

While waiting for these methods to produce reliable data, it is useful to study the *thermal conductivity* both experimentally and theoretically; a knowledge of it at depth provides valuable information even if it becomes confused through convection. The apparent conductivity, in addition to the usual phonon conductivity by vibrations of the crystal lattice, includes a radiative conductivity, silicates becoming transparent in the infrared to allow radiative transfer of heat, and a conductivity by excitons (transfer of an excited state from one atom to another). These last two components change very rapidly with temperature. Figure 94 (Lubimova, 1967) assigns an un-

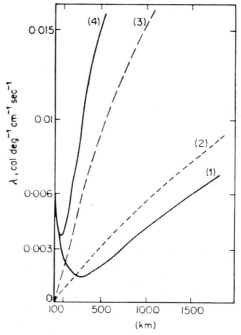

Fig. 94. Thermal conductivity with depth in the upper mantle: (1) phonon conductivity; (2) radiative; (3) exciton; (4) total. (Lubimova, 1967.)

usual share to exciton conductivity. The judgments involved here are really rather subjective; for example the radiative conductivity can be greatly modified by the presence or absence of secondary minerals opaque to the infrared. Qualitatively, the existence of a minimum near the surface, followed by a rapid rise, may be admitted, but if a vicious circle is to be avoided, it must be borne in mind that the choice of a relation between temperature and depth had to be made before the figure could be drawn. Lubimova regards the low conductivity of the uppermost layers, hindering the dissipation of heat to space, as an argument for progressive heating of the Earth's interior.

If no allowance is made for convection or shallow radioactive sources, the temperature gradient varies inversely with the conductivity; that is the temperature rises more and more rapidly down to the depth of minimum conductivity and thereafter less and less rapidly. In this way, the effect of temperature tending to reduce the rigidity might temporarily predominate over the effect of pressure tending to increase it. This would be a simple explanation for the low-velocity layer, but it would put the layer too far down. Accordingly we fall back on the usual interpretation in terms of proximity to melting. From this we should have most valuable evidence on temperatures in the mantle if we knew how the melting point varied with pressure or depth in the corresponding parts. As nothing much can be said about the melting of a body without knowing its composition, this point will be discussed first.

The only interpretation of the *continental* Mohorovicic discontinuity to stand the test of time is the one mentioned in Chapter VI, regarding it as a transition from basic rocks, basalts or gabbros, to ultra-basic rocks, peridotites and pyroxenes. Ringwood has suggested the name *pyrolite* (Clark and Ringwood, 1967; Mercy, 1967) for a rock of composition corresponding to one part (by weight) of basalt with three parts of dunite, the latter essentially constituted of olivine. Expressed as oxides, the proportions by weight would be 43% silica SiO_2; 39% magnesia MgO, 8% ferrous oxide FeO, ... This is roughly the composition of the silicate phase of chondrites, meteorites frequently considered to be representative of the Earth's mantle. The idea is that fractionation by melting of pyrolite would have produced the magma from which the basaltic layer derived, while the less easily melted dunite residue formed the part of the mantle just under the Moho. At the same time almost all the radioactive elements passed into the basaltic phase giving it an activity similar to that of present-day tholeiitic basalts.

This view of peridotites as residues neatly avoids the difficulties entailed in trying to derive the basalt from them. Although it is oversimplified and far from generally accepted we shall adhere to it. A petrographic verification would be difficult; pyrolite must take varied mineralogical forms depending on the temperature, pressure and presence or absence of water. Then too, its residue would have to be compared with those of terrestrial rocks considered to be samples of the upper mantle. According to Mercy these are neither the peridotite nodules in basalts, nor the peridotites of Alpine folds, but the garnet peridotite nodules which constitute the principal inclusions in diamond-bearing pipes; we cannot go into such details.

At depth the peridotite layer is supposed to end at the low-velocity layer. Under the oceans, according to the hypothesis of Hess, the serpentinized fraction of the peridotites lying above the Moho would form the oceanic layer; but the lithosphere would also include a lower non-serpentinized layer.

By assuming that pyrolite at atmospheric pressure is in first approximation composed of olivine, pyroxenes and plagioclase feldspars, and by making use of the known curves of equilibrium for simple systems, McConnell *et al.* (1967) have attempted to

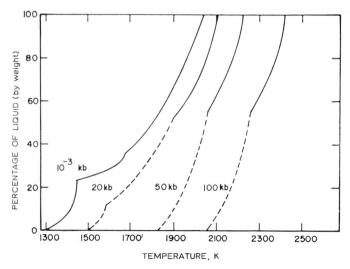

Fig. 95.　Melting curves for pyrolite at different pressures. (McConnell *et al.*, 1967.)

get an idea of the thermodynamic properties of pyrolite in its different forms, and particularly of its melting point. This is evidently not defined in the same way as for a pure material; it corresponds mainly to the melting of the lightest silicates, and may even be lower if eutectic mixtures are formed. The authors first determine the melting-point curve at atmospheric pressure (Figure 95); it is made up of several arcs; in the upper branch only the olivine solidifies, in the middle segment the pyroxenes, while olivine re-melts, finally in the lower segment the plagioclases. The curve is then modified to correspond to higher and higher pressures. At 100 kbars the plagioclases have disappeared. Figure 96 shows in a different form the domain of

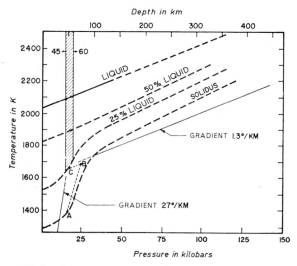

Fig. 96.　Melting possibilities of the upper mantle on the pyrolite hypothesis. (McConnell *et al.*, 1967.)

fusion of pyrolite in the diagram (pressure or depth, temperatuie). The most interesting curve is the 25 % curve corresponding to the complete melting of the basalt fraction, assumed to be in a constant proportion (of one part basalt to three of dunite), for all pressures. This curve and its neighbors have a rapidly rising portion which corresponds to the plagioclases and which disappears at high temperatures.

The authors assume the source of basaltic magma to lie between 45 and 60 km, the depth of Hawaiian earthquakes; this corresponds perhaps only to a lower limit of the actual depth (Kuno, 1967). They schematize the creation of basaltic magma around 50 km by the curve ABC of Figure 96: if the temperature happens to rise within an enclosure, the pressure goes up from A to B, with formation of a little liquid, which is sufficient to fracture the enclosure. The pressure then returns to normal along BC, with complete melting of the basalt which enters the fracture and produces an intrusion. The authors emphasize that their model guarantees a uniform composition foi the basaltic eruptions without postulating the existence of extensive liquid regions. The intrusions which cause the magnetic anomalies on the ridge crests are probably due to somewhat similar processes. This would be an important point for clarification.

In Figure 96 the two straight lines from C correspond to temperature gradients of 27°/km and 1.3°/km which the authors consider to be reasonable for the crust and the region of radiative conductivity respectively (convection undoubtedly lowers the second value). The two lines taken together in their view give a representation of mantle temperatures, or at any rate set an upper limit to them. Actually the explanation they give of volcanism, which we have just mentioned, would contradict the continued existence of a partially melted region. The general state of the mantle apart from volcanic regions might be represented by placing an intermediary segment between the two lines to follow, for example, the tangent to the curve of incipient melting (solidus) at A. Another region (near 100 km), in addition to the vicinity of A, then appears near the melting point; this would be the one corresponding to the low-velocity layer. In this way we should have the information we sought on the temperature; but its weakness is evident.

The chemical homogeneity of the mantle from the low-velocity layer down towards 900 km which would follow from the pyrolite hypothesis and which is essential for convective theories, is not in contradiction with the elastic heterogeneity that we have noted. In fact since the work of Bridgman we know how readily pressure can produce polymorphic transformations bringing minerals into denser and denser forms. Birch in 1952 attributed the apparent heterogeneity of the upper mantle to the superposition of a large number of these transformations, several of which are now well established (Clark and Ringwood, 1967; Anderson, 1967; Mercy, 1967). The simplest, suggested by Bernal as early as 1936, is that of orthorhombic olivine to a cubic crystal with spinel structure; others involve a pyroxene and an oxide, silica SiO_2 or magnesia MgO. In 1961 Stishov and Popova showed that quartz and its high pressure form coesite are transformed at around 73 kbars and 97.5 kbars respectively into a new dense form, having the rutile structure, subsequently named stishovite, and Birch suggested in 1964 that the deep mantle, elastically homogeneous as a result of all the polymorphic transformations, might best be pictured as a mixture of oxides in their densest form, the principal ones of course being SiO_2, MgO, and FeO. This rounds out a provisionally satisfactory sketch of the mantle as a whole.

Can the phase changes oppose convection by requiring the environment to provide the heat of transformation, restored, it is true, on the return journey? This question, complicated by the possible maintenance of metastable states, has been studied carefully by Verhoogen (1965). He concludes that at the pressures within the convective mantle the known transitions should be little hindrance. On the other hand, as Elsasser (1968) emphasizes, those transitions involving a volume increase on the rise and a decrease on the fall may take an important part in the operation of the convective thermal engine by amplifying the effect of the ordinary volume changes usually considered by the theoreticians. In 1968 Dennis and Walker also sought to attribute deep-focus earthquakes to the sudden contraction of metastable phases during the descent; but as Sykes has shown, the study of first motions is not favorable to this revival of an old theory.

Finally we note that phase transformations may play a role in the rapid increase of electrical conductivity with depth, to judge from the behavior of fayalite (the Fe_2SiO_4 constituent of olivine) whose conductivity rises a hundred-fold in changing to the spinel form (Akimoto and Fujisawa in 1965).

4. Rheology of the Mantle

The distinction that we have drawn between the lithosphere and the rest of the mantle may be convenient but remains artificial. There is no recognized discontinuity at the beginning of the low-velocity layer. Moreover an efficient coupling between the two is required by the mere fact that the mantle drags the lithosphere along. This raises the problem, unfortunately unsolved, of the laws of deformation of the matter in the Earth's interior. For purposes of calculation it is sometimes assumed that this matter responds elastically to stresses of short period like those of earthquakes, while following the laws of Newtonian viscosity for slow convective motion, in other words that it behaves as a visco-elastic Maxwellian body. It seems more natural to regard the mantle as a crystalline solid; in this case its flow should resemble the hot creep of metals in which lattice dislocations play an essential role. The laws for this case are quite different, but it is physically plausible that for very small stresses they should again reduce to a viscous deformation by migration of the dislocations. In the limit, as Tozer points out, solids, having a finite binding energy, flow under vanishing load by diffusion of atoms or ions, while the complications arising from grain boundaries and even dislocations come in only at temperatures far from the melting point and for large deformations. Unfortunately the longest possible experiments are too short to characterize clearly the corresponding domains.

We end by assuming that viscosity gives a generally correct representation of convective phenomena. But of course if viscosity alone were involved, there could be no understanding of any tectonics. On this point seismic evidence is instructive and we shall say a little about it.

The existence of visible faults proves that rocks are brittle under low pressure. A shallow-focus earthquake begins by an accumulation of more or less elastic stresses, then a rupture occurs at one point (focus) and propagates along the fault, whose movement reduces the stress. The mechanism of deep-focus earthquakes must be different, as Jeffreys already recognized in 1936. The hydrostatic pressure there may reach 200 kbars, which Orowan (1965) remarks is of the order of magnitude of the molecular cohesion between two welded solids; a shear stress of the same order would be needed for brittle fracture. Now according to Orowan the stresses causing an earthquake are at most of the order of 100 bars; his reckoning is open to criticism (Coulomb and Jobert, 1967) but there is no possibility of multiplying the result by a thousand. It will be assumed that beyond a certain depth dependent on the region but estimated by Orowan (1966) at less than 20 km, earthquakes result from a creep instability; the creep begins accidentally and is accelerated by the deformation itself producing disorder in the structure and raising the temperature locally. In this way the creep becomes concentrated in thin zones constituting faults of a sort.

Just where these phenomena occur evidently depends on the nature of the material. Elsasser as well as Orowan have emphasized the possible action of certain substances in making the lithosphere more plastic. The main agent is water; its effect on quartz is well known, and a few parts per hundred are enough to depress the melting point of basalt by 500 °C. As it begins to descend in trench regions, the material losing water would become brittle. In this line of thought the experiments of Rayleigh and Paterson are often cited. These showed that serpentine, which has a breaking strength comparable to that of granite at low temperature and high pressure, becomes brittle through certain dehydrations when the temperature rises. The curve of transformation, which occurs around 450 °C for iron-free serpentine, was determined by Scarfe and Wyllie (1967), but its exact relation with the evolution of the oceanic layer on Hess' hypothesis remains to be determined.

5. Convection by Vertical Instability

We have adopted the hypothesis of viscous convection. The problem immediately arises of estimating the value of the coefficient of viscosity at different depths. We shall set aside data on the quality factor Q as still poorly organized and hard to accommodate to slow phenomena, although much may be expected of it. This leaves us two pieces of information to work on. The first comes from the slow rising of the areas depressed by the weight of Quaternary glaciers and uncovered on their retreat, that is, Fenno-Scandia and the Canadian shield. The glacial trough, after a more or less elastic decompression of the lithosphere, has been filled in by a slow flow in the underlying regions, down to around 1000 km depth. The corresponding coefficient of viscosity is found to be of the order of 10^{21} or 10^{22} P. (See for example Coulomb and Jobert, 1963; Clark and Ringwood, 1967; Knopoff, 1967.)

The second is that observations of artificial satellites have proved that the equatorial bulge of the earth is too large to correspond to hydrostatic equilibrium at the present rate of rotation. Munk and MacDonald have ascribed this result to the probable decrease in the rate through tidal effects; the bulge would be partly fossil. A viscosity of 10^{26} or 10^{27} P for the Earth as a whole may be deduced from this, as MacDonald did in 1963 (McKenzie, 1967a).

With great reserve as to the physical significance of viscosity coefficients much higher than could be attributed to metals at ordinary temperatures, the two requirements expressed above may be satisfied simultaneously by assuming that the viscosity, very high in the lithosphere, falls to the order of 10^{21} P in the upper mantle, to rise rapidly again around 900 km and greatly exceed 10^{26} P.

Let us apply this to investigating the possibilities of convection, limiting ourselves for the moment to the convection generated by vertical temperature differences. Classic reasoning appeals to the theory of cellular convection in a viscous liquid, whose essentials date back to Rayleigh. He was concerned to explain Bénard's experiments on the formation of regular convection cells in a thin layer gently heated from below. The most perfect cells are hexagonal like the cells made by bees, with the liquid rising in the center and sinking along the walls; but many other arrangements are possible which depend on minute variations in the parameters of the system. In particular, when a slow horizontal motion is superimposed on the convection, rolls may occur, rotating in alternate senses aligned either parallel or perpendicular to the horizontal velocity; analogous motions in gases explain certain systems of parallel clouds (Coulomb and Loisel, 1940). Rayleigh showed that such stationary convections can exist *only if the heat sources keep the temperature gradient at a critical value*, always greater than the adiabatic gradient, and proportional to the viscosity. For a lower gradient the medium is in equilibrium; for a gradient a little higher the velocity of the currents increases with time; for a still higher gradient the motions become highly irregular. In all cases the system is unstable, and the location of the rising motions is determined by accidental circumstances.

This marginal character of cellular convection leaves small hope of getting a satisfactory model out of it readily. Nevertheless, application of the Rayleigh criterion to likely values of the temperature gradient and the viscosity shows (Knopoff, 1967) that the lower mantle must be sub-critical (no convection), and the upper mantle super-critical (irregularly convective), which agrees with our basic choice.

Tozer (1967) is the only theoretician to have approached the problem of viscous convection in the Earth with the generality desired. The essential features of his analysis are consideration of the enormous variation of viscosity with depth on one side, and consideration of the displacement of the heat sources by the convection itself on the other. Tozer's models are quasi-static, in other words their evolution is sufficiently slow for the composition of the matter, including the radioactive elements, to be considered invariable at each point, and for the same to be true of the temperatures at the boundaries of the layer. The next task, and a difficult one, is to define a starting model in hydrostatic equilibrium, with heat transmitted by conduction, from which to pass to the quasi-static model in treating the departures from it as infinitely small. This starting model or static model is necessarily spherically symmetric, so that convection resulting from horizontal differences in temperature would have to be excluded at this stage, if we had not already done so. However this does not keep Tozer from considering temperature conditions and the distribution of radioactive sources separately for the continental shields and for the oceans, using the data of Clark and Ringwood.

At the beginning of his investigation of the starting model, Tozer adopts an empirical formula, involving the melting point, for the variation of viscosity with temperature in the convective layer. At 600 km the viscosity jumps by four orders of magnitude. Calculation on the static model shows him that convection brings about an almost adiabatic distribution of temperature; Figure 97 shows the result. The temperature

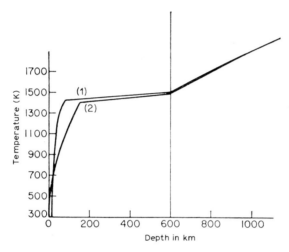

Fig. 97. Static solution for oceanic (1) and Precambrian (2) mantle. Convection ceases at 600 km because of a discontinuity in viscosity there. (Tozer, 1967.)

differences near the surface for shields and oceans are due to the higher radioactive concentration under continental regions. Convection begins at 50 km under the oceans, and it is there that the two temperatures differ most. Convection only begins at 100 km under the shields; from there on the difference in the two temperatures becomes insignificant.

The temperature distribution in the convective region is nearly insensitive to the strength of the heat sources; if the temperature rose, the viscosity would fall immediately, with convection becoming more intense and reestablishing equilibrium. This is an important result in view of our scanty knowledge of the internal radioactivity.

As well as the temperature, Tozer's static model determines the internal viscosity; convection maintains it at a nearly constant value, of the order of 10^{20} P, but farther down, under the convective layer, the viscosity changes abruptly to a magnitude of the order of 10^{25} P. Beyond this depth, still fixed at 600 km, heat exchange is by conduction, largely radiative. All this is in good agreement with all the views developed above, although the seismological evidence would call for the adoption of a somewhat thicker convective layer.

In passing from the static model to the definitive model, Tozer, although unable to make much theoretical progress, brings out some delicate distinctions. In addition to the possible irregularity of the currents due to crossing the critical Rayleigh gradient (an irregularity which has nothing to do with the appearance of mechanical turbulence in a laminar flow), there may be an irregularity due to a local excess of heat sources (the criteria are similar; instead of the heat flow entering the layer by conduction, the stationary flux generated there is considered). On the other hand Tozer cannot clarify some important details; for example the relative cross-sections of the rising and falling columns, or the spatio-temporal structure of the irregular convection.

Theory lacking, surprising results have been obtained under his direction in experiments on convection with internal heating of the liquid. One of the features observed is the descent of the fluid in the center of the cells, the opposite of what occurs in the ordinary Bénard cells. Another, more important, is that the horizontal dimensions of the cells are larger, for the same thickness, than in Bénard cells, which are little extended. Distances of several thousand kilometers between the rising and sinking limbs of a convective roll then seem compatible with a thickness of 600 or 800 km for the convective layer, removing an objection often made to theories of convection limited to the upper mantle.

For Tozer this convection bearing along its heat sources is characteristic of the oceans. Under the continents, the upper mantle, impoverished by the concentration of radioactivity in the crust, would be more suited to convection of the Bénard-Rayleigh type; whether or not it exists is of little concern for our subject. But we must reflect on what happens under the continental margins.

6. Convection by Horizontal Instability

Hitherto we have confined ourselves to convection involving a vertical instability. As early as 1935, Pekeris had considered the effect of horizontal differences of temperature between continents and oceans, but he assumed convection of the entire mantle. In this case again it is much more realistic to suppose it limited in depth.

An essential difference with Rayleigh convection is that the system is unstable in the presence of a horizontal temperature gradient, no matter how slight. Now all calculations of interior temperatures based on heat flow values at the surface and on the assumed radioactivity of the oceanic and continental crusts lead to large temperature differences throughout the low-velocity layer; the temperature under the oceans would exceed that under the continents by a hundred degrees or so: Knopoff (1967) therefore assumes that convection occurs all along the continental margins in this low-velocity layer (Figure 98). The velocity of the current decreases very rapidly with the thickness of the convective layer. Allan, Thompson and Weiss (1967) have shown, however, that for a simplified model involving a kinematic viscosity (viscosity divided

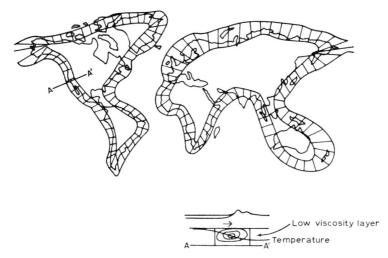

Fig. 98.   Schematic view of the marginal zones where convection by horizontal temperature gradients may be expected. The section shows higher temperatures in the oceanic mantle than under the continent. (Knopoff, 1967.)

by density) of 10^{21} stokes down to 500 km, 10^{26} beyond this, velocities of the order of 1 cm/y. should be produced over a half wavelength of 3000 km. In these circumstances the horizontal instability by itself seems to explain sea floor spreading. But this prospect appears to us to go against the fact already mentioned (Figure 97) that ordinary convection considerably reduces the temperature difference between the continental and oceanic mantles.

In a more modest way, horizontal instability perhaps allows some understanding of the peculiarities presented by island arcs and epicontinental seas bounded by them. Indeed the model of Oliver and Isacks (Chapter IV) is hardly applicable in cases where a second trench is located behind the one in which the lithosphere is supposed to be engulfed. In Japan, for example, the crust does not go down with the lithosphere; the Moho rises to form a classic root under the archipelago (Japanese national report for the Upper mantle project, Tokyo, 1967). Another example: behind the Tonga trench lies the plateau of the Fijis, and beyond that again, the New Hebrides trench, with a focal surface sloping the other way. In addition (Chapter VII), the heat flow in epicontinental seas or in the Fiji plateau is close to the mean flow observed on the ridges (with less dispersion in the values, no doubt because of their smoother relief); Oxburgh and Turcotte (1968) have attempted to explain this by the bold hypothesis of friction at the region where the lithosphere descends. Finally the form itself of island arcs may be compared to those forms like scales assumed by convection cells when placed in a general current, well known from the example of cloud cells (Figure 99).

The epicontinental seas are thus to be distinguished from other margins. They alone could be the seat of a local convective circulation modified by the general current of expansion, the boundary between the two circuits being the focal surface of deep

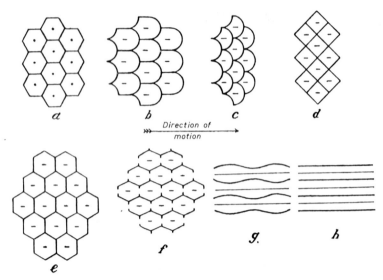

Fig. 99. Development of convection cells in a weak but increasing current. (A. Graham; Figure
from J. Coulomb and J. Loisel, 1940.)

earthquakes. As support for these evidently speculative views may be adduced; (1)
the odd shape of this surface under Japan or in the Sea of Okhotsk, perhaps even in
the Tongas, where it seems to flatten out at around 200 km in the one case (Figure 4),
around 350 km in the other (Figure 5); (2) the existence of lineated magnetic anomalies
(Chapter II) in the Sea of Okhotsk, and of anomalous mantle (Chapter IV) in the
Sea of Japan; (3) finally the bringing to light of a kind of ridge inside the South
Sandwich arc* by Griffiths, Ashcroft, Barker and Parkinson (communication at
Zurich, 1967).

Any explanation of the oceanic arcs should also be applicable to terrestrial arcs
like the Himalayas and Carpathians. Hess (1965) postulates a ridge in the middle of
the Tethys at the beginning of the Mesozoic and Glangeaud has found evidence of
it in the western Mediterranean. In correlation with this, Hess compares serpentine
dredged in the Pacific to Alpine ophiolites. Finally, it has been seen in Chapter VII
that heat flow is high in the Carpathian basin. The possibility of marginal convections
thus seems to deserve careful examination.

* More recently, Menard, et al. (1968) find linear rises and deep narrow trenches near the Fijis,
which are seismic, and resemble a center of expansion. Shor et al. (1968) find an uplifted oceanic
crust there.

CHAPTER IX

OCEANS AND CONTINENTS. CONCLUSION

1. Birth of the Oceanic Crust

We have appealed to highly diverse branches of geophysics. We might have gone further. For example, meteorites may teach something about terrestrial chemistry; or again the mantle temperature depends on the way heat escapes from the core, and the theory of the terrestrial dynamo might be able to furnish information in this regard. But the benefits are lost in the growth of the margin of uncertainty. It is time to conclude by returning gradually to the observable phenomena and by describing some features which have hitherto not found a place. We shall dispense with repetition of the arguments already given in the preceding chapters in support of each assertion.

The expansion observed at the surface reflects, as we have seen, the motion of a roughly symmetric pair of convective rolls, much broader than deep, rising under their common partition located somewhere under the ridges. The perfect symmetry of the ridges was explained in Chapter V. The stagnant zone lying over the partition must be where differentiation of the anomalous mantle occurs. Perhaps the latter should be regarded as the plagioclase variety of pyrolite which is stable at low pressure, although requiring high temperatures. The stagnant zone is responsible for the flaring shape of the base of the anomalous mantle far above it, a shape suggested by Figure 72 (case 1), caused by the progressive separation of the two rising currents.

The anomalous mantle operates as a physico-chemical factory whose task is to produce and emit as follows: upwards, the basement basalts as intrusions (coming to the surface if the ridge has an axial valley, otherwise not, but in either case strongly magnetizable) and as ordinary volcanic eruptions; sideways, the peridotites of the normal mantle; wedged between the two, the serpentinized peridotites of the oceanic layer, derived from the preceding products by an addition of water which it seemed must be juvenile (Hess, 1962), although isotopic analysis appears to call for a descent of sea water. These transformations, whose exact site and petrographic nature remain indefinite, taken together would cause a large increase in volume raising the ridge (the density of olivine being 3.3 g/cm^2 and of serpentine 2.6, 70% serpentinization over 5 km accounts for only 750 m).

One of Hess' main arguments in favor of serpentinized peridotites is that the thickness of the oceanic layer is found to be fairly constant at around 5 km, which he explains by a fairly constant temperature of hydration of about 500°; if 2 km of basement are added to 5 km of oceanic layer, the hypothesis implies a gradient of 70°/km and a heat flow of 3 μ cal/cm^2 s (assuming a conductivity of 0.004 cal/cm s °C, of

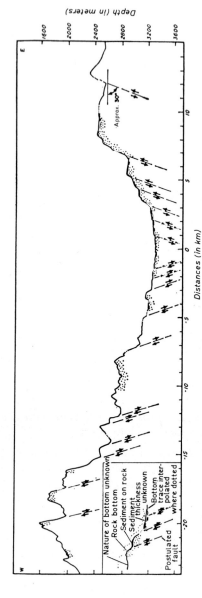

Fig. 100.   Gorda ridge. Profile of the median valley from vertical and oblique sounders towed near the bottom (vertical exaggeration 5 times). (Atwater and Mudie, 1968.)

the order of that of basalt). These are very reasonable figures under the crestal region of ridges. During the spreading, assumed to be uniform, the cooled and rigid layer would retain its composition. However, its thickness increases somewhat with distance from the axis, which presumably calls for some modifications to the picture.

The oceanic layer, which is continuous under the East Pacific ridge, is absent under the crest of the Atlantic ridge (Figures 66 and 15). For Cann (1968) this absence would be understandable only if the 500° isotherm was located too high in the crust; whereas the Atlantic heat flow is lower than the East Pacific flow and the isotherm is lower. From what we know of the physics and chemistry of the anomalous mantle, this objection does not seem to condemn the hypothesis of Hess outright; it is less serious than the want of continuity between the oceanic Moho and the continental Moho. However Cann, with the support of the inverse correlation observed by Le Pichon *et al.* (1965) between the thicknesses of the oceanic layer and the basement under the mid-Atlantic ridge, proposes the ingenious solution already mentioned in Chapter VI, in which the oceanic layer would be constituted from top to bottom from the basement basalts, metamorphosed into greenschists and then into amphibolites. The difficulties of this model, which evidently preserves the duality of the Mohos, seem as great as those of Hess' model.

It would be desirable at this point to explain in detail how the basalt intrusions responsible for the magnetic anomalies rise from the basement, and what their relation is to the median valley when the spreading rate is low. Some clues have been provided by an original study of the Gorda ridge by means of a group of instruments (magnetometer and various sounders) perfected at the Scripps Institution and towed a few hundred meters above the bottom (Atwater and Mudie, 1968; communications at Zurich of Spiess, Mudie and Harrison, and Harrison and Mudie). The axial valley shows a stepped structure which may be interpreted as the normal faults of a graben (Figure 100). The steps lean outward and carry perched sediments which Atwater and Mudie assume to be turbidites arriving constantly from the continent into the bottom of the valley from its southern end. The same tipped block structure, masked by sediments and volcanic outflows, remains perceptible up to 90 km from the axis.

In addition, by observing near the bottom, the aligned magnetic anomalies are again found, but very much stronger; they show a fine structure corresponding to wavelengths of the order of 3 km and amplitudes reaching 2000 γ, probably due to dykes extruded in the vicinity of the axis which come to pollute the regions of reversed polarity, as Matthews imagines (Chapter III). Unfortunately, the correlation between the magnetic anomalies and the faulted topography is not always visible.

2. Death of the Oceanic Crust

What happens at the other end of the convective system? Let us first consider the case of zones of descending lithosphere in regions of trenches and deep-focus earthquakes (Pacific rim, Antilles, South Sandwich Islands). Even if Elsasser's suggestion of the lithosphere sinking of its own weight is rejected, we have seen that it acts like a rigid plate over reasonable distances. Thus it is not absolutely necessary to suppose that the horizontal motion of the convective roll accompanies the lithosphere all the way, with an inclined current still pushing it in its descent. It is more difficult to situate the convective descent when the lithosphere has not met with an obstacle obliging it to flex; for example under North America where the Atlantic and Pacific currents converge. Perhaps the convective descent there is broad and diffuse, or consists of two distinct zones for the two currents, or is even inexistent; in this last case there would be 'open cell' convection, as Orowan maintains (1965). The mobility of the lithospheric plates on the surface masks the deep currents and complicates the old idea of each continent coming to rest directly over a descending partition.

We could take advantage of our ignorance to draw a single line of descent, continuous and closed, topologically similar to the seam of a tennis ball (Figure 89); this would teach us little.

In the zones where the lithosphere penetrates the upper mantle it presumably remains identifiable down to the limit of deep-focus earthquakes, that is for 1000 km at a slope of 45°. If the spreading rate is 5 cm/y., for example, the descent lasts 20 m.y., during which the lithosphere may experience physico-chemical transformations. Some of those which occurred on the rise are reversible, but the conditions are not symmetrical. In particular, Hess (1962) looks into the fate of the serpentine. It would lose its water as soon as it reached 500° again. This water, thrown out by volcanoes, would according to him increase the volume of the oceans. Hess thinks that all the water of the sea and atmosphere might have originated in this way. Here is his calculation scarcely modified: a modest 30000 km of ridges emitting at the low rate of 1 cm/y. a 5 km thick layer serpentinized to 70% and so containing 25% water by volume, would furnish 0.4 km^3/y., or 1.6×10^9 km^3 in 4000 m.y., of which 0.3 is still in the oceanic layer. Thus the present volume of the oceans is arrived at, say about 1.3×10^9 km^3. The dehydrated peridotites would not be put back into circulation, but would be added to the continental mantle, and thanks to this the continents would not be submerged by the rising water, acting together with erosion.

Dietz (1966), who does not accept the gradual constitution of the hydrosphere, but who also is interested in erosion, offers some much more detailed ideas on the growth of continents, ideas which raise implicitly the problem of the differentiation of granite. Dietz in effect takes up the case where the floor spreading comes up against a coast, actually the Atlantic coast of the United States. According to the rigid plate model, for example in Le Pichon's synthesis (Chapter IV), nothing remarkable is presently going on there. But Dietz aims to explain the Appalachian foldings. It may be supposed that when they were being uplifted, America offered a greater resistance to the spreading of the Atlantic than at present (Wilson, 1968 has even defined some periods when the Atlantic Ocean is supposed to have reclosed). This would lead to the establishment at that epoch of a trench structure, like that of present day South America. However nothing analogous to Andean volcanism will be seen in Dietz' mechanism.

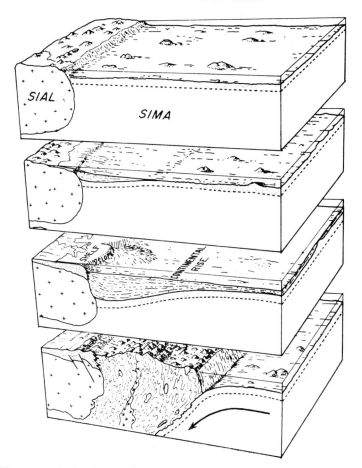

Fig. 101. The geosynclinal cycle and the growth of continents. The prism of sediment on the pre-continental rise is transformed into folded mountains of graywacke or flysch type. (Dietz, 1966.)

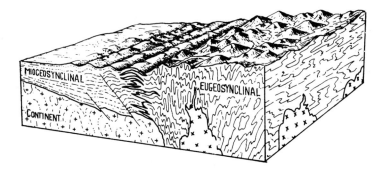

Fig. 102. Part of the present Appalachians. The crystalline Appalachians *(right)* are strongly deformed and invaded by plutons. Their complexity results from sea-floor movements. The folded Appalachians *(left)* rest on the continent, whose immobility preserved them from intense orogeny.
(Dietz, 1966.)

Dietz allots the basic role to sediments, coming from the continent or swept from the oceanic floors and accumulating in the precontinental rise, which would slowly sink under the thrust of the lithosphere. For a somewhat obscure reason, which ought to be deducible from the sinking itself by use of theories of the origin of granite by transformation of sediments (maintained in France by Wyart), the mass is invaded by plutons and uplifted (Figure 101). It rises onto the continental shelf, which remains passive, and induces some foldings there by gravity. A trench develops at this stage, but the main result, made explicit in Figure 102, is the recurrence of the distinction between eugeosynclinal (the crystalline Appalachians, coming directly from the precontinental rise) and miogeosynclinal (the folded Appalachians, resting on the old continent).

Dietz, as others have done, believes that continents grow by annular zones. The foregoing mechanism would correspond to the formation of a part of a ring; why the ring should be complete is not too evident. The continents are thus supposed to grow without deformation, except at the edges, and to maintain their level isostatically in spite of erosion.

Dietz' views, with no place for any but marginal orogenies, and too categorical as to the annular distribution, were criticized in 1967 by Young (Dietz, 1966, discussion); Young maintains, in particular, that the Appalachians, and the Caledonian chain, originated from an 'intracratonic' geosynclinal in the midst of a massive continent. The question is related to the reconstruction of the continents before the opening of the Atlantic. Dietz and Sproll (Falcon, 1967) favor two separate primitive continents (Laurasia and Gondwana), and so Dietz does not allow Africa and the present United States to have been in contact; besides this contact is a weak point in Bullard's reconstruction (Chapter V). In his reply to Young, Dietz finally admits that certain geosynclinals are indeed born in the middle of stabilized continents, but maintains his description for those showing the eugeosynclinal-miogeosynclinal pair. He recalls that in eugeosynclinal ranges, such as the Coast Range of California, enclaves of serpentines and of ultrabasic rocks are found, which would come from the oceanic layer and the upper mantle respectively; he might have mentioned like Hess the ophiolites of the Alps.

Without wishing to bring into question again this agreement of a sort among geologists for a double origin of mountain ranges, and while seeing few difficulties for metamorphism and volcanism in the interior of a continent, the author of the present book admits never having clearly understood how the other classic episodes of geosynclinal orogeny could have taken place there. The birth of ranges on the edge of a sea traversed by a ridge and closing up, like the Atlantic according to Wilson (1968) or the Mediterranean according to Glangeaud, seems much clearer to him.

3. The Ocean Floor as a Conveyor Belt

We have seen that the ocean floor was renewed in a time of the order of magnitude of 100 m.y. This floor has been compared to a conveyor belt, and we should like to conclude by examining what it is really capable of transporting besides the film of basalt responsible for the magnetic anomalies.

Sediments and various rocks, first of all. We have alluded to this several times, and made use of the information obtained from it to construct spreading rate scales. A detailed discussion of the observations would be tiresome, but a rapid survey will be conclusive.

On the whole the unconsolidated sediments of the Atlantic ridge date from the Pleistocene or later. This is the rule in the crestal zone; however sediments or rocks may be retained within the transverse fractures: blocks of basalt in the Atlantic fracture near the St. Paul Rocks, containing more or less indurated sediments with microfossils of the Lower Miocene (Saito *et al.*, 1966); Paleocene foraminifers also coming from fractures near the St. Paul Rocks, identified by Cifelli, Blow, Melson (Phillips, 1967); discordant Pleistocene sediments on horizontal deposits of Upper Eocene age in the Vema fracture. On the flanks, cores frequently penetrate sediments earlier than the Pleistocene, the oldest being a calcareous mud of the Upper Cretaceous found over 3 m of red clay at 1170 km from the axis, itself lying on serpentine pebbles indicative of the basement (Saito *et al.*, 1966).

The Lamont team has found similar results on the East Pacific rise (Burckle *et al.*, 1967). On the crest the cores run into rock or penetrate one or two meters of Quaternary sediments. On departing from it, the age of the oldest sediment in each core increases. The Pliocene is reached on the upper flanks, Upper and Middle Miocene at mid flank, and Lower Miocene on the outer flanks. An Oligocene core was secured at 1600 km from the crest, then some Eocene cores. However, no Cretaceous sediments have been found east of the ridge, while they are found to the west, from the central Pacific on. The results of the Scripps Institution team (Harrison and Mudie, 1967; Riedel, 1967) in the Clarion and Clipperton fracture regions are a little less clear.

After sediments, seamounts. Dymond and Windom (1968) have found for three of these, lying southwest of Hawaii, minimum ages of 85 to 90 m.y., in good agreement with the Lamont magnetic age for the region which is 90 m.y. If the peaks were born on the ridge axis and have never lagged behind the corresponding anomaly, the mean spreading rate since the Cretaceous has therefore not exceeded the recent rate on which was based the extrapolation of Heirtzler *et al.* (1968). One of the peaks is thought to show recent volcanism (0.7 m.y.). Therefore conditions at depth during the displacement of the lithosphere remain favorable for magma production.

There are still not very many ages measured by the potassium-argon method, and much use has been made of guyots, truncated and submerged ancient volcanoes discovered and named by Hess in 1946, by estimating the epoch at which their tops were planed down by the waves.

The normal life sequence for a guyot may be envisaged, after Hess (1965), as follows: its *birth* on the crest of a ridge; its *departure* from the ridge, borne along by the lithosphere until its volcanism ceases and its upper platform is created, which may take a few million years; its *submersion* through its descent on the flanks at a vertical rate of the order of 30 m/m.y. if the spreading rate is 1 cm/y. (this rate allows the

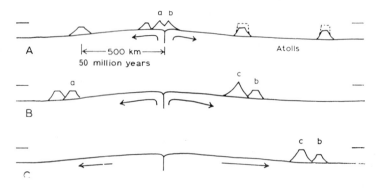

Fig. 103. Formation of guyots and atolls. At time A volcanoes a and b form on either side of the axis. At B they have become separated to form a symmetric pair of guyots; volcano c develops off the axis. By time C, c has been truncated, has descended the slope and been submerged, but its top rises above that of b. (Hess, 1965.)

corals to keep building if the water temperature is suitable, in which case the end result is an atoll instead of a guyot); finally, the *disappearance* of the atoll or guyot in a trench, as Menard and Dietz have been maintaining since 1952 of the guyots between Alaska and the Aleutians.

Save for Vogt and Ostenso (1967), hardly anyone has sought to check these ideas against active ridges, although some guyots or atolls are thought to be related to the Atlantic, Indian or East Pacific ridges. Menard (1965) in fact finds that the correlation between submarine volcanoes and ridges only shows up well for the Darwin rise; however he defines its limits very liberally. Menard thinks the other ridges have not yet reached the age for being covered with volcanoes, which would be a sign of their decline.

According to Hess, there is an approximate concordance in the depths of the summits of the guyots in the region of the western Pacific where they abound. This seems to him consonant with the preceding mechanism; here and there, a probably younger guyot formed on the flanks (Figure 103), has a few hundred meters less depth.

However other phenomena may account for the submersion of guyots: a eustatic

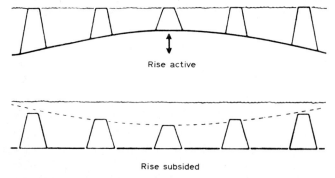

Fig. 104. Depth of guyots on the Darwin rise. (Menard, 1965.)

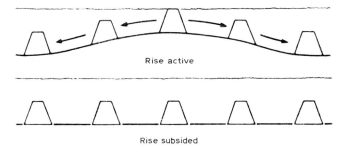

Fig. 105.   Depth of guyots on the Darwin rise on the hypothesis of Hess. (Menard, 1965.)

variation of sea level, an individual puncture of the crust by the guyot, or lastly the slumping of an inactive ridge. The first two processes can scarcely have the necessary amplitude. If the last is given the preference, and it is assumed that the guyots grew at random on the flanks and were rapidly eroded, their present depth should be greatest on the former crest (Figure 104), which is what is observed according to Menard (1965b) for the whole Pacific, in opposition to the relatively constant depth given by Hess' mechanism (Figure 105).

Although Menard says nothing about it, this reasoning would lead primarily to the supposition that the Darwin rise has never been a ridge in the proper sense. It is possible however that some guyots ascribed by Menard to the Darwin rise come from the East Pacific, such as those of the Tuamotus, according to Vogt and Ostenso (1967).

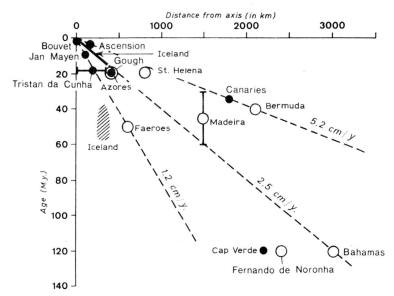

Fig. 106.   Age of oldest rock in various Atlantic islands, as a function of distance from the ridge axis. *Solid circles* denote the presence of active volcanoes, the *bars* show uncertainty of age or distance. (Vogt and Ostenso, 1967.)

Let us come lastly to the larger islands. Wilson (1965) has listed the maximum ages of the islands of the Atlantic. The deviations which may be noticed in Figure 106 are probably explained by the differences (with distance along the ridge and with the passage of time) in the spreading rate, or its angle to the ridge; by the possible discovery of older rocks on the islands; by the possibility of their generation off the axis. Continuance of volcanic activity during the displacement of the lithosphere is evident from the presence of active volcanoes in the Cape Verde and Canary Islands, well beyond the ridge, on the outer edge of the corresponding basins (one might wonder whether the islands have been temporarily submerged in them).

Thus nothing would be left behind by the sea floor conveyor, not even the fissures or magmatic chimneys, rising in all likelihood from the low-velocity layer, which come through the lithosphere and continue their role during its displacement.

4. Conclusion

We are now on the point of concluding. The present generation is reaping the harvest of half a century of exploration of the oceans. But here, as in other fields, discoveries come more and more rapidly. Vening Meinesz believed that he had found the key to orogeny in the island arcs. Thanks to Ewing, geophysicists turned their interest to the ridges. The ideas of Hess, the explanation of magnetic lineations by Vine and Matthews have led to linkage of the two structures through sea floor spreading. The recently discovered transverse faults have found their place in the system.

We have not disguised the difficulties of detail which explanations still run into; it may even be that we have overemphasized them, at the risk of allowing the great overall success to be forgotten. The main problems still to be resolved concern the physics and chemistry at depth on one side, and on the other, the way in which the convective currents determine and modify the position of the ridges. It would also be important for our understanding of orogeny to know whether the surface structure reacts appreciably on the convection itself. Great efforts of observation and of imaginative theory are still necessary, but a long step forward has been made. The next will be toward the depths.

CHAPTER X

RECENT PROGRESS

1. Introduction

To cover in a single chapter the important developments since the French edition of the present book, we must confine ourselves to results with wide implications. The concept of an only slightly deformable, mobile plate which appears more and more essential (Isacks *et al.*, 1968), has become an indispensable tool of tectonic thinking, although it requires caution in application to small scale systems. Geophysicists are making way for geologists, and regional studies are becoming so numerous that they can only be mentioned as examples.

In the matter of organization, the international project of Upper Mantle study in the framework of the International Union of Geodesy and Geophysics, which inspired much research in our field of interest, has just ended. A new organization, named Inter-Union Geodynamics Commission, formed jointly by the Union of Geodesy and Geophysics, and the Union of Geological Sciences, seeks to investigate the deep-lying mechanical and thermodynamic causes of surface movements. Everyone is fully aware that the plate theory cannot explain everything, despite its remarkable success in the oceans. The Geodynamics Program reminds us that the basic contrast between oceanic and continental crust leaves room for intermediate cases (most continental margins); for explaining the origin of folded ranges and oceanic basins, it is important to know whether some of these cases correspond to a material transformation from one crust to the other. Apart from this, the principal deficiencies of the theory are in dealing with vertical movements, major foldings (the Rockies), and earthquakes within the continents; vertical movements in the oceans, well observed recently on the ridges (Figure 108) and on their transverse fractures, are also concerned. Eruptive activity within oceanic plates (Hawaii, guyot clusters) or within continental plates (Deccan basalts), and the existence of aseismic rises (Walvis) are not well understood either.

Although new results have come from many sea expeditions, special interest attaches to the JOIDES operation, Joint Oceanographic Institutions for Deep Earth Sampling (Hammond, 1970). This was a series of campaigns of deep sea drilling carried out by the ship *Glomar Challenger* for a consortium of universities and research institutes with the financial support of the National Science Foundation. Its aim was to date sea floors in all regions where the potassium-argon method fails to provide accurate ages. This required cores reaching the basement, whose age is then known as directly antecedent to the stratigraphic age of the sediment in contact with it. These operations were divided into a series of legs begun in August 1968. The first 13 legs comprising 219 drillings, some of which were of 1 km depth under 6 km of water, include 5 in the North and South Atlantic, 5 in the Pacific, 2 in the Gulf of Mexico and the Caribbean, and one in the Mediterranean. In addition to various articles, three volumes of reports (Initial reports of the deep sea drilling projects, U.S. Government Printing Office, Washington, D.C.) have now been published. Already it may be inferred that the previously established history of sea floor spreading is perfectly applicable to the South Atlantic, and with some complications, almost equally so to the Pacific. For want of competence we shall be brief on the results concerning sediments; however the new ideas of Dewey and Bird as to the origin of the rocks constituting emerged chains will be examined.

In terminating this introduction, I should like to thank all those who have supplied me with information, particularly Claude Allègre, Jean Francheteau, Kurt Lambeck and Xavier Le Pichon, and those who have given me permission to use their illustrations.

2. Magnetic Anomalies and the Sea Floor

De Boer *et al.* (1969) appear to have verified reversal of magnetization on pillow lavas dredged from the axis of the Reykjanes ridge and on either side.

Mapping of magnetic anomalies in the oceans has been extended (Vacquier in P. J. Hart, Ed., 1969; Heirtzler, *ibid.*; Vogt and Ostenso, 1970; Schlich and Patriat, 1971, etc.). A curious example is the Great Magnetic Bight in the Gulf of Alaska discovered by Elvers, Matthewson, Kohler and Moses in 1967, and mentioned in Chapter II. Ages of the anomalies, which increase from east to west off America, increase from north to south near the Aleutians, that is, in departing from the trench, cut diagonally by one or two anomalies. The main cause of this complex phenomenon is the birth of the Aleutian arc, probably in the Eocene (Grim and Erickson, 1969; Jones 1971). Another interesting case is the region off Lower California (Taylor *et al.*, 1971), whose complications are likely due to the displacement of North America since the Miocene; spreading rates along the ridge are observed to vary abruptly in places, smoothly in others (between 2.07 cm/y. and 4.24 cm/y. over 5° of latitude).

Rona *et al.* (1970) have found ancient linear anomalies, within the magnetic smooth zone (Chapter II) of the region between the Canaries and the Cape Verde Islands, which are symmetric with the anomalies of the American continental margin. The interpretation of the two magnetically quiet zones is still under discussion, however (Vogt *et al.*, 1970).

The structure of the anomalies observed close to the bottom (Chapter IX) has been clarified by Luyendyk (1969), and by Larson and Spiess (1969). The latter have observed the first three magnetic reversals in the region where the crest of the East Pacific rise enters the Gulf of California. They arrive at an upper limit of 4700 y. for the duration of a reversal, and at 280 m for the width of the basaltic axial intrusion, while recognizing an occasional break of a few kilometers in the intrusion. Short period fluctuations in magnetization might be due to the terrestrial dynamo.

Attention may be drawn to the appearance of two new methods of dating anomalies in addition to the potassium-argon method with improved details (Evans, 1970) and the micropaleontology of ocean sediments: fission tracks method applicable to basaltic glasses (Luyendyk and Fischer, 1969) have furnished an age determination of 35 ± 5 m.y., beyond the range of K-Ar; Bada *et al.* (1970) have used the progressive decline in the rotatory power of the amino-acids in organic sediments to obtain the sedimentation rate and the age (1.23 m.y. at the base) for a 518 cm core, which unfortunately had no recognizable reversal. The direct reversal scale seems firmly established except for minor events (Emilia and Heinrichs, 1969; Evans, 1970; Denham and Cox, 1970) whose date, and more particularly duration, remain uncertain. This scale has even been used to check the dating of recent sediments (Ku *et al.*, 1968).

Beyond 5 m.y. the anomaly age scale now depends on the JOIDES results. These have confirmed (Heirtzler, 1970) the Lamont scale to -30 m.y., and even from -30 to -80 with small discrepancies, while beyond -80 m.y. ages remain uncertain. However, the resolution of magnetic measurements at sea is no better than 50000 y. The world-wide halt of Ewing and Ewing (1967) has not been rediscovered.

An idea of the scatter in the results will be seen in the Atlantic around $30\,^\circ$S (Maxwell *et al.*, 1970):

Distances from the axis (km)	191 ± 5	380 ± 10	643 ± 20	727 ± 10	990 ± 10	1270 ± 10
Paleontological ages of sediments in contact (m.y.)	11 ± 1	24 ± 1	33 ± 2	40 ± 1.5	49 ± 1	67 ± 1
Magnetic ages from Dickson *et al.* (1968)	9	21	34–38	38–39	53	70–72

The spreading rate has remained near 2 cm/y. without any long-period variations.

The youthfulness of the sea floor in all oceans is now definitely recognized, as may be seen in Figure 107 for the North Pacific (Fischer *et al.*, 1970). The age of the oldest known oceanic sediments, found in the North Atlantic just beyond the continental margin of America, is 165 m.y. (Middle Jurassic); this implies the existence of an ocean there in the Mesozoic, with a spreading rate twice the present one. Perhaps older basement might have been found in the North Pacific if the drilling had not encountered layers of chert. These layers are also present at places in the North Atlantic, with ages of 70 m.y., where they serve as important seismic 'markers'. (Ewing *et al.*, 1970).

Finally, JOIDES drillings have shown that sea floors have undergone large and rapid vertical displacements, in addition to horizontal ones. The lower sediments of certain cores contain limestone while the upper sediments lack it; they were laid down therefore below the level of solution. This could come about through the descent of

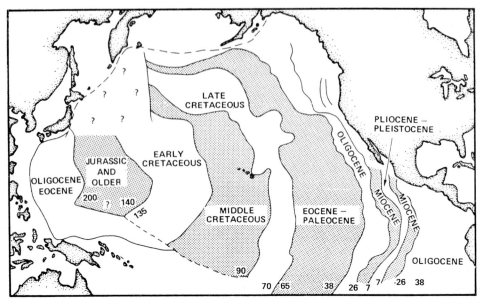

Fig. 107.   Age of the oceanic crust in the North Pacific. (Fischer and Heezen, 1971.)

the floor on the ridge flanks (Fischer *et al.*, 1970); in the northern-most Atlantic however, fragments separated from the continents during the expansion are thought to have been engulfed very rapidly. Full interpretation of the sedimentation observed in the South Atlantic (Figure 108) on the other hand, calls for variations in the level of the ridge itself, perhaps accompanied by short term fluctuations in the spreading rate (Maxwell *et al.*, 1970).

The anisotropy of the lithosphere below the sea floor due to sea-floor spreading (Chapter VI) has been confirmed (Morris *et al.*, 1969). The fact of its reaching 8% (against a maximum of 19% possible for a single crystal) suggests the possibility of a slow recrystallization under tension.

3. Ridges and Rifts, Transform Faults

Numerous studies have added to our knowledge of most of the noteworthy lines of seismicity. We shall analyse the major ones, first those involving boundaries in tension and then those under compression. The great slip faults, particularly the San Andreas, have been investigated, with a view to forecasting earthquakes, by a detailed study of the stresses they give rise to, or the way in which the displacements are apportioned between abrupt breaks and gradual slippage, but we shall not deal with them.

We begin with the ridges. Van Andel and Ross Heath (1970) reveal the existence of distant faults parallel to the mid-Atlantic ridge.

Weertman (1971) makes unexpected use of the theory of elastic dislocations in

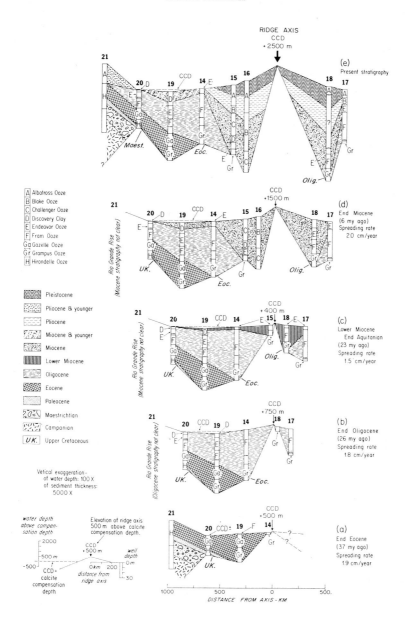

Fig. 108. Reconstruction of the history of the South Atlantic, with sea-floor spreading, paleontological ages, and lithologic formations taken into consideration. The distance between sites, the bathymetry, and the stratigraphical thickness are shown at different scales for effective illustration. The top line connecting each section represents the past topography; the bottom line illustrates the sediment-basalt contact. (a) End of Eocene (37 m.y. ago); spreading rate, 1.9 cm per year. (b) End of Oligocene (26 m.y. ago); spreading rate, 1.8 cm per year. (c) Lower Miocene to end of Aquitanian (23 m.y. ago); spreading rate, 1.5 cm per year. (d) End of Miocene (6 m.y. ago); spreading rate, 2.0 cm per year. (e) Present stratigraphy. (Maxwell *et al.*, 1970; courtesy *AAAS*.)

applying it to a mechanism for the ascent of lava: the lava collecting under a plate in tension opens a crack which increases in volume, pinches off at the base, rises to the surface as if seeking hydrostatic equilibrium, then solidifies and blocks further extrusions.

Some results of a geographic nature may be mentioned: the west branch of the Indian Ocean ridge, often considered inactive, is instead badly fractured and has a low spreading rate of 0.6 to 0.9 cm/y. (Schlich and Patriat, 1971). The Chile ridge, topographically similar to the mid-Atlantic ridge, with a spreading rate of 2 to 3 cm/y., is moving away from the East Pacific rise (Herron and Hayes, 1969) but the evolution indicated by the magnetic anomalies is complex.

The relative simplicity of cases where only oceanic crust is involved contrasts with the much greater complexity of emerged structures. This is already apparent where the mid-Atlantic ridge goes through Iceland (Ward et al., 1969; Serson et al., 1968), and even more so in the continental grabens. The area of the Red Sea, Gulf of Aden and the African rifts is still the best investigated of these (Falcon et al., 1970; McKenzie et al., 1970); the Afar depression gives an idea of the complexity of a triple ridge junction (Tazieff, 1971).

The area of the western United States which continues the East Pacific rise (Chapters II and IV) is being extended by normal faulting. Elsasser (1971) regards this as a necking in the thickness of the continental lithosphere corresponding to an extension at depth; the rupture now evidenced by the Gulf of California might thus move northward. Artemjev and Artyushkov (1971) interpret the formation of the Baikal graben also by a plastic necking of deep layers followed by subsidence of the rigid upper layers, the only ones subject to normal faulting. As the crust is thin under the graben, the negative Bouguer anomaly would be caused by the density deficiency of the overlying sediments, unlike the model of Figure 73.

The ways in which oceanic crust and lithosphere are formed and differentiated within ridges have not been explained completely by the many petrographic studies made. Details will be found in the forthcoming publication of a Royal Society meeting (November 1969) on 'The Petrology of Igneous and Metamorphic Rocks from the Ocean Floor' (Phil. Trans. Roy. Soc. A268, no. 1192, 365 p.). Let us recall the basic difference between the tholeiites (K_2O around 0.1 to 0.2%; Na_2O around 2 to 3%) emitted on the crests and representing 80% of oceanic volcanic rocks, and the alkaline basalts (K_2O around 1 to 2% with minor CaO and Na_2O content) which constitute, with the tholeiites, the volcanic islands and submarine mountains. According to Kay et al. (1970), the tholeiites originate in the partial (around 30%) melting of mantle peridotites, followed by a fractionation of the olivine and plagioclases through crystallization at shallow depth (15 to 25 km).

In ocean fracture zones the tholeiites are often metamorphosed and demagnetized. Dredgings from deep fractures such as the Romanche trench (Melson and Thompson, 1970) also fetch up coarse basic rocks (gabbros, diorites) and ultrabasic rocks, especially serpentines (Chapter VI), all more or less altered and metamorphosed. It is still a question whether the latter are representative of the upper mantle, constituted perhaps of lherzolites (Engel and Fisher, 1969) or of garnet pyroxenites (Reid and Frey, 1971). Miyashiro *et al.* (1970) suggest two possible models for ridge petro-graphy, drawn from those of Hess and of Cann respectively (Chapter IX). They prefer the second, which does not invoke large amounts of juvenile water, whose practical absence at the Earth's surface is shown by isotopic analysis of H or O. In this model, the fracture zone serpentines are supposed to be formed *locally* by hydra-tion of peridotitic intrusions. The crust would be of metamorphosed (and so de-magnetized) basalts, except for the thin surface layer carrying the anomalies; it is interesting that the difference between layer 2 (basement) and 3 (oceanic layer) is supposed to be mechanical in nature, with the highly fractured layer 2 becoming consolidated in layer 3. The anomalous mantle would consist partly of plagioclase peridotites, partly of amphibole peridotites, also fractured.

Models for the oceanic lithosphere have been sought in such ophiolitic complexes as that of Troodos in Cyprus, supposed to be derived from it. It may also be mentioned that Medaris and Dott (1970) think that a group of peridotites found in Oregon, cut by small gabbro dikes, might be ancient ridge material transported by sea floor spreading from the Gorda ridge.

Ridge investigations must be linked to the study of oceanic transform faults, generally readily recognizable by their bordering crests (up to 2 km high), by offset of the magnetic anomalies and so on. Recently studied fractures include the Endeavour, southwest of New Zealand (Christoffel and Ross, 1970), and the Gibbs, between the mid-Atlantic ridge and the margin of Labrador (Olivet *et al.*, 1970). In the equatorial Atlantic the prolongation of the Romanche fracture zone has been traced under the sediments of the Gulf of Guinea (Fail *et al.*, 1970).

4. Compressive Boundaries

In the oceans the lithosphere coming out of the ridges sinks into the asthenosphere after a horizontal travel of sufficient length, according to Dewey and Bird (1970), to cool it and give it a greater density than the asthenosphere (Elsasser, 1968; see Chapter VI). Dewey and Bird see in the length of this travel the difference between the Atlantic and the Pacific; the Antilles, however, are near the mid-Atlantic ridge, and Mexico is even closer to the East Pacific rise!

What finally becomes of the downgoing matter? Elsasser (1970) notes that basalt production on the ridges (6 km^3/y. according to Menard, 1967b) is six times the amount needed to produce the present crust during the age of the Earth; return of the basalt to the asthenosphere would provide a reservoir of material for epeirogenic uplift.

The downgoing slab of lithosphere is often called a Benioff zone, although Guten-berg had long before recognized the corresponding disposition of earthquake foci. These occur however only over a thickness of perhaps 5 km; 25 km at most for the intermediate depth Tonga earthquakes (according to Mitronovas *et al.*, 1969). The slab is roughly flat and of rather variable dip, (according to Luyendyk, 1970, inversely correlated with the distance from the center of rotation); this is still ill-explained theoretically. The delay in seismic wave travel times through this slab has been well observed; by Davies and McKenzie in 1969; by Toksöz *et al.* (1971) as well as by Dubois (1969) under the New Hebrides arc. Similar delays under New Caledonia lead Dubois to think it may have belonged to an island arc extinct in the Oligocene (however see Gardner, 1970).

Stress distribution in the plunging slab is known from many studies of the focal mechanism. A survey of Isacks and Molnar (1969) deals with 14 fairly plane seismic

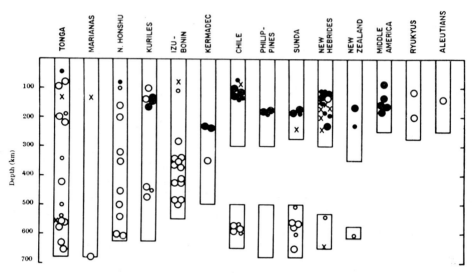

Fig. 109.   Down-dip stress type plotted as a function of depth for fourteen regions. A filled circle represents an orientation such that the axis of tension is within 20–30 ° of the local dip of the zone; that is, down-dip extension. An unfilled circle represents an orientation such that the axis of compression is within 20–30 ° of the local dip of the zone; that is, down-dip compression. The X's represent orientations that satisfy neither of the preceding. Smaller symbols represent less reliable determinations. The enclosed rectangular areas approximately indicate the distribution of earthquakes as a function of depth by showing the maximum depths and the presence of gaps for the various zones. The zones are grouped (from left to right) according to whether the zone is continuous to depths of 500–700 km, discontinuous with a gap between intermediate depth and deep earthquakes, or continuous but reaching depths less than 300–400 km. (Isacks et Molnar, 1970; courtesy *Nature*.)

zones (Figure 109). It is generally found that either the axis of maximum compression, or the axis of minimum compression (axis of tension) is directed approximately along the line of steepest slope of the plane. (The axes are determined on the assumption that the nodal planes separating compressions and dilatations are planes of maximum shear.) The fault plane itself is never parallel to the seismic zone plane. The earthquakes thus do not appear to be attributable to frictional forces between the downgoing slab and the asthenosphere, which would be small in any event, but rather to ruptures within the slab itself. Near the surface it is drawn out by its own weight (Elsasser, 1968, see Chapter VI); further down it is compressed on meeting with the resistance of the lower part of the asthenosphere; this effect propagates back up the slab to put it all under compression. In some cases, however, gaps are observed in the distribution of foci with depth; this is because the slab has broken under traction before reaching the zone of compression. The region where rupture occurs, coinciding with the 'intermediate' earthquakes limit of Chapter I, may be connected with phase transitions further increasing the density of the cooled lithosphere; in strongly arcuate regions the gap may disappear at the end of the arc (Figure 110).

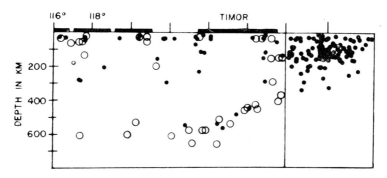

Fig. 110. Projection of the seismicity onto a vertical cylinder parallel to the Sunda arc. Open circles represent the best earthquake locations computed by the USCGS for the years from 1961 through 1968, closed circles others locations computed by the USCGS for this period. In the Banda Sea region (east of 128° E) the good locations were not separated from the others. (Fitch and Molnar 1970; courtesy *Am. Geophys. Union*.)

Shallow focus earthquakes, in ocean trench regions where the sediments are rarely folded (Chapter VI), usually correspond to a more or less horizontal extension, thought by Isacks *et al.* (1968) to be caused by the bending of the lithosphere. According to Malahoff (1970) normal faults inside the trench would conform farther down with the direction of descent of the slab; but fractures perpendicular to the arc are also observed.

The arc shape might be attributed to the fact that the lithosphere goes down first where it is most cooled; but no general process is discernible to explain the occurrence of double arcs, sedimentary and volcanic, with sometimes a third arc, such as the Aves ridge in the Antilles (Karig, 1970). Nor do petrographers agree on how the andesitic lavas characteristic of volcanic arcs are produced (Montigny *et al.*, 1969).

We finish with oceanic arcs by mentioning the interesting case of the two volcanic arcs, north of Celebes, and on Halmahera, studied by Fitch in 1970. Between their opposing convexities, intermediate focus earthquakes lie on two planes forming an inverted V without any horizontal lithosphere between them; the region is noted, since the studies of Vening Meinesz, for its gravity anomalies, which are biconcave.

Continental compressive boundaries (McKenzie, 1969) present a more complex case than the island arcs, and this is not merely because of a wealth of geological knowledge. It is because the continental plates, although thick, are often dislocated and reconstituted, and so less resistant than the basic or ultrabasic layers of oceanic plates. Their rocks are also lighter, at least the granite, so that continental lithosphere cannot sink, and consequently earthquakes in it are almost exclusively of shallow focus. Their seismic zones, which surround the aseismic shields, appear to be diffuse, but many epicenters may be related to overthrustings, including those located behind the arcuate ranges forming the South Eurasian boundary (Fitch, 1970). These arcuate ranges have no andesitic volcanoes, but exceptionally deep focus earthquakes are located in their vicinity. McKenzie supposes them to correspond to a remnant of an oceanic slab whose upper part has disappeared while its lower part continues down under its own weight. Probably this clustering of continental deep foci in strongly curved regions (Hindu Kush, Carpathian Bight, Betic cordillera) is to be connected with the slowing down of the descending slab in the similar oceanic regions, as it appeared in the preceding figure.

5. Kinematics of Plate Motions

Plate motions, treated by McKenzie (1970), have been determined for a great variety of regions. In straight-forward cases concordant results are obtained by using spreading rates on ridge axes (more than a hundred are now known according to Le Pichon), the geometry of transform faults (which are within 5 or 10 km of small circles centered on the pole of rotation), or the focal mechanism of earthquakes, from which can be deduced the direction of movement and even some statistical indications as to its amount (Brune in 1968; Northrop et al., 1970). New arguments for the plate idea, for any who may still have doubts, are furnished by the propagation of the S_n seismic phase. This is a transverse wave travelling between the Moho and the low velocity layer, which readily passes through stable regions (shields, basins), even when they are cut by transform faults, but only with difficulty through ridge crests and the interiors of arcs, showing that the lithosphere is discontinuous in these regions (Molnar and Oliver, 1969).

Changes in direction of the fractures lead to a first complication, for when a ridge changes direction in the course of emission, the flow lines are deflected. The faults, unable to follow the curvature, readjust by local rupture. When a rapid change in direction involves a very long ridge segment, the latter is broken up in the creation of new transform faults, which may later disappear, or may develop further. Rea (1970), Francheteau et al. (1970) have investigated the extreme case of the fossil fractures of the East Pacific. Francheteau et al. define, for each of the great fractures, five successive matching segments. The first three segments correspond to magnetic anomalies with numbers 10–16, 16–25 and 25–32. The relative poles of rotation obtained from each group of segments are poorly defined in the direction perpendicular to the fractures; but the magnetic anomalies of the first three groups allow one to calculate the normal component of the spreading rate for the time when these anomalies were close to the ridge. It is known (Chapter IV) that this rate varies with the sine of the colatitude and hence the pole position can be determined. The pole moved at least 30° over the whole growth period of the fractures. Stability in the relative rotation of two plates tangent to a long fault is thus less than is commonly thought, as we shall see when we come to the reconstruction of ancient positions.

McKenzie and Morgan (1969) have studied the configuration of plate boundaries in the neighborhood of a triple point common to three plates, A, B and C. Whether the configuration can be maintained depends on the global interactions between plates, but there are local conditions which may be determined by assuming a plane region and rectilinear boundaries. Once established, the junction of three ridges (Figure 111) may persist, whatever the global movement is. This is evident if the

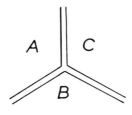

Fig. 111.

expansion is normal to the ridges; in the general case, conditions are needed for compatibility. On the other hand consider a junction of three arcs at a point O_1. In Figure 112, the sides with arrows (indicating the motion relative to the adjoining plate) are overriding edges and consequently attached to the corresponding plates. The boundary (B, C) thus remains fixed with respect to C; O_1 goes to O_2 on the new boundaries of A parallel to the former ones. The new boundary between A and B is now composed of two segments with the new part O_2O_2' in line with the (C, A) boundary. A point such as P_1 on (C, A) has arrived at P_2 on (A, B); for an observer attached to this point, the direction and magnitude of the overriding have changed, giving him the impression of a major change in the expansion.

McKenzie and Morgan show that transition to a configuration in which two of the boundaries are in line is the normal evolution of a triple point, except when the three boundaries are ridges, a stable situation as we have seen. They apply their theory to several cases; in particular they try to reconstruct the development of the complex structures indicated by the magnetic anomalies off central California.

Some authors have relied blindly on McKenzie and Morgan's results in spite of their local character, others have not chosen to take account of such considerations. This is the case with Krause and Watkins (1970) in an otherwise carefully executed study of the Azores triple point. Wishing to account for (1) the seismic fracture zone east of the Azores with a tension earthquake at its west end (2) the aseismic fracture zone west of the Azores (3) the alignment of the islands (4) the abrupt change of

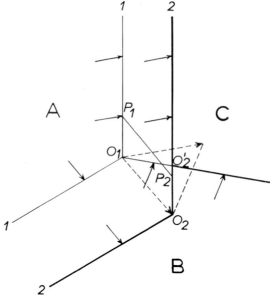

Fig. 112.

direction of the ridge, and its widening, they end up with the 'simplified' model of Figure 113, where the dotted strips indicate the magnetic anomalies and the x's the seismic fractures. This is a foretaste of the complications the theory of sea floor spreading will have to go into if it is to accommodate the complexity of the data. Along this southern boundary of the Eurasian plate, Le Pichon's model (1968) shows further that the rate of compression varies greatly between the Azores and the eastern Mediterranean; it is one of the regions to be studied in the Geodynamics project.

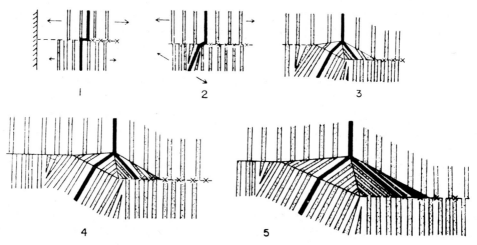

Fig. 113.

6. Dynamics and Thermodynamics of Plate Motions

The distinction between lithosphere and asthenosphere is not a difference in composition, as it probably is for the distinction between crust and mantle, but a difference in viscosity, although this is not a proper term if the relative motions are of the nature of non-linear creep, as metal studies seem to indicate (Weertman, 1970). Whatever type of flow is involved is very greatly facilitated by the increase of temperature with depth, so much so that for Elsasser (1971), horizontal shear is not transmitted from lithosphere to mantle or vice versa. In his view, the lithosphere is a 'stress guide', like a sheet of paper placed on a basin of mercury, which may be pushed or pulled practically without deformation. Of course there may be interaction between lithosphere and asthenosphere by heat exchange; the descending slab chills its surroundings and thus induces convective motions on them.

The asthenosphere is often assimilated with the low velocity layer (Chapter VIII) in spite of a ratio of 10^{12} (Elsasser) in the velocities involved. Little is known however of the physical conditions in it. Lliboutry imagines it to be at the melting point throughout like a temperate glacier. Nur (1971) regards it as a solid sprinkled with melted inclusions; the velocity and damping of seismic waves then depend on their frequency; at the moment there is no evidence for or against this. Lambert and Wyllie have followed Orowan in emphasizing the importance of plasticization by water, which hydrates the rocks and lowers melting points. The low velocity layer would be where fusion was beginning due to traces of water, and it would be water rising from the deep mantle which triggers the ascent of the basaltic magma (Wyllie, 1971). The number of variables is too large however to convince us to adopt his detailed models of the upper mantle.

The depth of the asthenosphere has been the object of much research. Kanamori and Press (1970) constrain models of the mantle under the oceans arrived at by a Monte Carlo method to give group velocities for surface waves agreeing with observation; they find that the velocity of S waves must decrease rapidly below 70 km (and return to normal around 300 km depth).

Walcott (1970) revives the model of an elastic plate resting on a fluid. Its behavior depends on its resistance to flexure $D = ET^3/12(1-\sigma^2)$ where T is the plate thickness, E Young's modulus, and σ Poisson's coefficient. D may be determined, though not very precisely, in regions burdened with sediments, or unburdened by erosion, the retreat of glaciers or simply by the lowering of a water level, as in Lake Bonneville (Utah), already studied by Crittenden, where old beaches have become dome shaped. The scatter in the values of D found by Walcott may be due to variations in the thickness T, but the values seem to correlate more particularly with the time taken by the plate to yield to the stress. Allowing for this, a thickness of the order of 110 km for the lithosphere is found, except in the Bonneville region where the thickness comes to the same order as that of the crust (20 or 30 km). This contrast calls to mind that in this Basin and Range province there is a seismic ridge covered over by the displacement of North America. Walcott's results, primarily applicable to the continental lithosphere, are not incompatible with a thickness of 70 km for the oceanic lithosphere.

Let us now consider the practically horizontal movement of the lithosphere, without Elsasser's assumption that it is necessarily decoupled from the movement of the asthenosphere. Lliboutry (1971) obtains some clues as to the nature of the return current by simply writing down equations for equilibrium. He considers a lithosphere and asthenosphere with uniform thicknesses h', h, and densities ρ', ρ. The asthenosphere also has a uniform viscosity. The viscosity below the asthenosphere, in the 'mesosphere' is very much higher, although its value is presently most uncertain; Goldreich and Toomre in 1969 have in fact been critical of its estimation from the slowing of the Earth's rotation. Lliboutry's results are summarized in Figure 114.

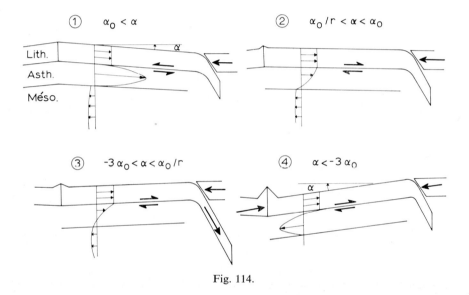

Fig. 114.

In it, $\alpha_0 = 2U\eta/\rho g h^2$, $r = 1 + (2\rho'h'/\rho h)$, with U being the velocity of the plate with respect to the mesosphere. α_0 is on the order of 10^{-4}. If the mesosphere is rigid, with the return taking place in the asthenosphere, α is negative (model 4). If the asthenosphere drags the plate along, with the return occurring in the mesosphere as in a classical convection cell, α is positive (model 1). The plate may drag the asthenosphere as it descends either under its own weight (model 2), or drawn down by the downgoing slab (model 3, or model 4 with rigid mesosphere). In this last case, α may be negative. Negative values of α seem to be excluded by the slope of the bottom, but if ρ' increases as the lithosphere advances, the slope of the isobars, actual factor in the equilibrium, may be opposite to that of the bottom. Jacoby (1970) has made similar calculations taking account of the wedge shape given to the lithosphere above the ridges by the ascent of the asthenosphere.

Further progress requires a choice between possible causes of the motion. Convection in a fluid asthenosphere may act on the lithosphere either by viscous drag or by uplifting the ridge regions, the plates then descending by gravity. McKenzie (1969), using the horizontal temperature gradient convection model of Allan *et al.* (1967), finds the first process to be more important in all likelihood. The second assumes a relation between the rates of descent and production of the lithosphere which may not always obtain under varying conditions.

Almost every possible form of thermal convection has been considered, even including (Shaw, 1970) the energy of tidal friction (which could be of local importance, although only contributing a few per cent at most). Even within the limits of cellular convection it would be impossible to describe all the models considered, stationary and non-stationary, dissipative and non-dissipative, with or without transport of heat sources, with base temperatures ranging from 500° to 1500°C, and so on. In each case also, an estimate is needed of the stresses transmitted to the lithosphere in order to know if they can displace or break it. Elder (communication, First European Earth and Planetary Physics Colloquium, Reading, 1971) for example, associates transverse fractures with secondary convective rolls elongated in the direction of the the principal current (Figure 99) and turning alternately in opposite directions. The convection observed in the relatively simple case of the large furnaces used in glass making (Peychès and Zortea, 1971) displays in fact a remarkable variety of current flows.

On the theoretical level, resort is had to numerical methods (Torrance and Turcotte, 1971) to take into consideration the variation of viscosity or of the creep parameters with temperature and pressure, but present computer capabilities limit such calculations to two dimensional models. Phase changes (Chapter VIII) must also play a role. Schubert, Turcotte and Oxburgh have shown that they may contribute to instability, even in those cases (such as the olivine-spinel transformation, or partial melting near the surface) where the light phase lies above the heavy phase, provided that the temperature gradient is high enough, as seems to be the case in the mantle (Schubert and Turcotte, 1971).

Convection in a closed circuit is certainly only a rough approximation; a very irregular pattern for the rise of matter can be imagined. Morgan (1970) assumes that there are hot spots in the asthenosphere, away from the ridges, giving rise to narrow plumes of matter. The passage of a plate over a hot spot results in an upflow of basalt unlike the tholeiitic basalt of the ridges, thus explaining the formation of island chains and aseismic rises. The three chains Hawaii-Emperor sea-mounts, the Tuamotus, the Austral rise-Gilberts-Marshalls, are supposed to have come from the successive passage of the Pacific plate over three fixed hot spots.

This contrasts with the formation of the volcanoes of arcs and ridges. The reservoir for the arc volcanoes, to be taken as fixed, is still less than 60 km down, as judged by the damping of S waves observed by Gorshkov in 1968 in Kamchatka, and by Matumoto and Molnar in 1969 at Katmai (Alaska). Lithospheric transport of ridge volcanoes has been confirmed by Menard (1969). He shows that recent crust in the East Pacific carries only small structures, which will become islands and then guyots (which are only observed on the oldest parts of the crust, where the sea floor is 500 to 1000 m lower than it was when they were islands).

It may be said in terminating this digression that such questions of transport are much more complicated for large islands like the Canaries for example. Dietz and Sproll (1970) regard the eastern islands (Fuerteventura and Lanzarote) as fragments of the African continent, the rocks of the others being typically oceanic. Bosshard and Macfarlane (1970) conclude from the seismic and gravimetric data that the others are independent volcanic structures resulting from eruptions along NE-SW fractures, suggestive of secondary ridges.

If Elsasser's views are followed, the motion of plates must be ascribed primarily to the drag of the cold slab of lithosphere on them. Let us consider this hypothesis more closely. McKenzie (1969) examines the temperature distribution in the down-going slab, a problem mathematically similar to his treatment of ridges (Chapter VII). Reheating of the slab is fairly slow. Starting with oceanic lithosphere 50 km thick with a temperature distribution between $0\,°C$ and $T\,°C$ at constant gradient, and assuming the mantle to be at a constant $T°$, McKenzie finds that a reasonable choice of parameters leads to temperatures in the middle of the slab of under $0.6\ T$ down to nearly 300 km, $0.8\ T$ near 600 km, for a slab velocity of 10 cm/y. He adopts $T=800\,°C$ and endeavors to show, using the seismic belt Tonga-Fiji-Kermadec-New Zealand as example, that earthquakes cease beyond $0.85\ T=680°$. Their greatest depth is thus proportional to the down-going velocity. Similar calculations, greatly elaborated, have been carried out by Toksöz et al. (1971) in an effort to deduce the behavior of seismic rays in traversing the lithospheric slab, a question which had been taken up by Davies and McKenzie in 1969.

Knowing the temperature of the slab and thus its density (in the absence of phase changes), McKenzie determines its drag on the lithosphere, finding a considerable $2.5 \sin \varphi$ kbars, where φ is the dip of the slab. McKenzie next assumes a Newtonian viscosity of $\eta = 3 \times 10^{21}$ P for the asthenosphere, given by postglacial uplift, and determines the motion of the lithosphere in the asthenosphere, the corresponding shear σ, which exceed a hundred bars on each side of the descending slab, and the frictional heating σ^2/η. This is given in units of 10^{-7} erg cm^{-3} s^{-1} in Figure 115, but the uncertainty of it must be borne in mind.

Various objections have been made to Elsasser's mechanism, some by McKenzie himself. How did the motion start, and how is it transmitted from plates with a down-going slab to others without? Why should the African and Antarctic plates for example separate? It is surprising that the lithosphere supports tension; McKenzie replies that separation would require one of the principal pressures to be negative, which is impossible at depth; this argument is valid for a brittle solid but not very convincing here. Finally McKenzie compares the tension with the friction to be overcome, which depends on the horizontal length of the plate. He finds a critical length, somewhat similar to the situation in Rayleigh convection, greater than which nothing would move, but short of which all the plates would be displaced at the same speed; this is hardly in accord with observation.

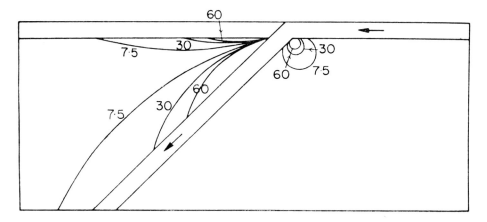

Fig. 115.   Stress heating caused by viscous dissipation in units of 10^{-7} erg cm^{-3} s^{-1}. The heating within the mantle is more intense behind the arc than in front of it. (McKenzie, 1969;
courtesy *Roy. Astron. Soc.*)

Let us in concluding draw attention to theoretical and experimental models of
Elder (1967) and Malkus (1971) in which the continents are self-propelled, moved
by the convective systems they themselves give rise to; through their horizontal
variations in thermal conductivity in Elder's model, or their horizontal variations in radioactivity in Malkus'.

7. Terrestrial Heat Flow

There has been a proliferation of heat flow measurements in the oceans. Here are
some results, of increasing complexity. Epp *et al.* (1970) find an excessive flux
(2.1 μ cal/cm^2 s) over the region of the diapirs of the Gulf of Mexico, where the
ground is highly conductive. They remark that the circulation of water near the bottom
or in the superficial sediments, and bacterial production of heat are perhaps not
negligible. Lister (1970) finds a smooth decrease in heat flow with departure from the
crest of the Juan de Fuca ridge, save for an anomalous station (volcanic?) at 60 km
distance with a doubled flow. Von Herzen *et al.* (1970) find a greatly variable distribution west of the mid-Atlantic ridge, but systematic in the sense of high values
within 200 km of the crest, and low values beyond 300 km, with abnormal values
in the Vema fracture zone attributable to tectonic and intrusive activity. More
surprising are the results of Talwani *et al.*, (1971) on the Reykjanes ridge. Heat flow
seems low up to about 10 km from the crest, and is certainly low beyond 50 km,
so that *the observed mean out to 100 km would be no greater than for the nearby ocean
basins!*

Sclater and Francheteau (1970) have introduced a little order into oceanic heat
flow measurements with the help of the magnetic anomalies and JOIDES drilling.
They distinguish nine provinces of different ages in the North Pacific. Mean heat

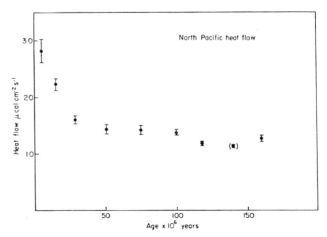

Fig. 116.   Plot of mean heat flow against age of province for the North Pacific. The length of the bar gives the magnitude of the respective standard error. The mean value for the youngest province has been plotted for a mean age of 5 m.y. in order to account for the paucity of observations in the crestal regions. (Sclater and Francheteau, 1970; courtesy *Roy. Astron. Soc.*)

flow decreases with age (Figure 116), which is explained (Chapter VII) by the cooling of the plate as it moves away from the ridge. There is a similar but less exact decrease in the South Atlantic. A decrease had also been found by Polyak and Smirnov in 1968 for continental orogenies of increasing age (Figure 117), but at one tenth the oceanic rate. This would result in part from the dissipation (in 200 m.y. at least, 400 m.y. at most) of the heat brought up by plutonic intrusions, in part from the

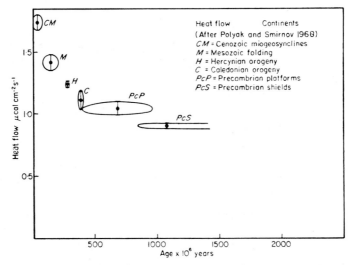

Fig. 117.   A plot of mean heat flow against age of orogenic province for continents, after Polyak and Smirnov, 1968. (Sclater and Francheteau, 1970; courtesy *Roy. Astron. Soc.*)

decrease in the radioactivity, particularly of potassium, of the intrusives. The heat flow at depth would remain nearly constant.

Equality of heat flow (at 1.1 μ cal/cm^2 s) over shields and over old basins holds fairly well if shield values are increased a little to allow for possible errors in laboratory measures of conductivity. But this equality, and the same holds for mean flows, is merely the result of an accidental compensation between an excess of radioactivity under the continents and a diminution in the deep heat flow. Sclater and Francheteau show that equality is obtained in two plausible models; one having the same temperature at the base of the continental lithosphere as at the base of the much shallower oceanic lithosphere; the other having the same heat flow on the two bases. Both models assume an oceanic plate thickness of 75 to 100 km, with the continental plate almost twice as thick. In both models the temperature around 100 km depth is markedly higher under the oceans.

8. Epicontinental Seas

Epicontinental seas, new examples of which have been studied (Milsom, 1970), continue to pose interesting problems (Chapter VIII). The underlying crust is youthful; JOIDES drillings have shown (Figure 107) a transition from Jurassic crust to a crust only 30 to 40 m.y. old in crossing the Marianas trench, and that a similar transition occurs in crossing the Carolines arc. It is also to be recalled that heat flow in these seas is high (Chapter VII). In McKenzie's (1969) overly abstract model, the fluid asthenosphere is drawn to the interior of the arc by the movement of the descending slab (Figure 115); the friction of this fluid on the base of the lithosphere is supposed to furnish the excess heat. Let us examine the facts.

Dewey and Bird (1970) consider the Sea of Japan, where the Yamato bank has the appearance of a microcontinent, to be either a part of the Pacific cut off by the arc and accumulating sediments since the Triassic, or, on the contrary, to be the result of a recent detachment of Japan from Asia. The second hypothesis is much more likely. Indeed a detailed study by Karig (1970) of the Tonga-Fiji region suggests that the sedimentary and volcanic arcs advancing on the ocean leave behind them a basin with an undulated oceanic crust (Menard, 1964) and a high heat flow, sometimes bounded by a fixed 'third arc' (observation of the third arc dates from Vening Meinesz). There would thus be a one-sided sea floor spreading from pseudo-ridges and creation of new crust. In this connection Baranzagi and Isacks (1970) find a region of high damping for seismic waves, hence a gap or a constriction by necking in the lithosphere, between the volcanic arc of the Tongas and the Lau ridge, Karig's third arc. The Tyrrhenian Sea also (Heezen *et al.*, 1971) shows valleys and rises probably resulting from intrusions or volcanism along parallel fissures. It is well known as a region of subsidence, so that similar features might be expected in other subsidence basins, oceanic or continental. But there may also exist subsident basins with a continental crust thinned by the scouring of mobile underlayers ('subcrustal erosion'); it appears that the Rockall plateau, studied by Le Pichon *et al.* (1970), for which sinking has been confirmed by JOIDES drilling, may be of this type without spreading or new crust formation.

Oxburgh and Turcotte (1971) also explain sea-floor spreading behind arcs by intrusions (mainly granites or granodiorites). 4 μ cal/cm^2 s on 500 km depth in the interior of the arc would correspond to a geologically acceptable spreading rate of 5 cm/y. According to their model the Sea of Japan grows at 2.5 cm/y., markedly less than the rate at which lithosphere is being engulfed in the Japanese trench (order of 10 cm/y.).

9. Gravity Anomalies and Plate Theory

Although gravity observations have played only a secondary role in the development of plate theory, the effect of ridges and arcs on the global free-air or isostatic anomaly fields must be recognized before deep seated phenomena can be understood.

The anomalies over island arcs, discovered by Vening Meinesz, consist of a narrow negative band and a much wider but less intense positive anomaly in the concavity of the arc; the distance from the low to the high is about 115 km in the 7 cases studied by Hatherton (1969). Another extensive positive anomaly is often found on the outside of the arc.

Free-air anomalies over ridges are highly variable because of the uneven relief. On the average, Jacoby (1970) finds a weak high over the North Atlantic ridge and the East Pacific rise, a weak low over the South Atlantic ridge. The North Atlantic high appears clearly on a map of Talwani and Le Pichon (in Hart (ed.), 1969), but the situation is more complex in the South Atlantic.

The broad features of world gravity are provided by the study of satellite orbits, the details by measurements on land or sea. As each improved world gravity map appears, plate boundaries are looked for on it. The latest examination is by Kaula (1970, 1971), utilizing a development of the gravity potential, due to Gaposchkin and Lambeck, as yet unpublished, in spherical harmonics up to at least the sixteenth degree and order. Kaula subtracts the harmonics up to degree 5, which probably originate at depth, and considers isostatic anomalies (Figure 118). Most ridges are seen to be associated with positive anomalies, but paradoxically, the same is true of arcuate regions.

It might be thought that in a regime of slow flow a regional mass excess could be produced by the substitution at depth of dense materials for less dense ones (in the case of arcs), or by the arrival of excess material at the surface (ridge case). But Kaula notes that the effect depends on the boundary conditions. Warm, light material, for example, rising under a fixed boundary causes a negative anomaly; a free boundary rises, producing a positive anomaly, masking the first. The same boundary must be considered as fixed or free depending on its time of reaction to the developing stresses. Thus the lithosphere is taken to be free under the ridges, rigid on the exterior of arcs, and often on the interior (Andes; continental arcs). The resolution (1200 km) of the development is not sufficient to show Vening Meinesz' bands of negative anomalies.

Anomalies are negative over ocean basins, the shields, and Antarctica.

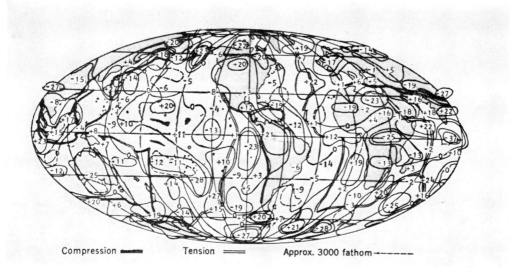

Fig. 118. Isostatic anomalies in milligals referred to a 5th degree spherical harmonic figure. They are calculated from the spherical harmonic coefficients of the gravitational field of degrees 6 to 16 of Gaposchkin and Lambeck and from the coefficients of the Airy-Heiskanen isostatic reduction for a 30-km nominal crustal thickness by Uotila. (Kaula, 1970; courtesy *AAAS.*)

10. Orogenesis on the Plate Theory

Dewey and Bird (1970) have attempted a general explanation of mountain range formation using the idea of repeated opening and closing of the oceans, as upheld by Wilson for the Atlantic or by Glangeaud for the Mediterranean. It is a considerable improvement of the views of Dietz on the role of precontinental rise sediments (Chapter IX); unfortunately our presentation must be brief.

Dewey and Bird adopt a ridge model in which, except for the alkaline volcanism of the flanks, layer 2 is formed of tholeiites, layer 3 of dikes in its upper part and layered gabbros in the lower. Under the Moho the transition is to ultrabasic rocks such as harzburgites and lherzolites. The rocks of layer 3 are supposed to show alterations induced from above in the ridge region in accordance with Cann's view (formation of zeolite, chlorite, amphibole from basalts and gabbros), or from the mantle below as Hess would have it (injection of serpentine, particularly along the parts of transverse fractures near the ridge where the throw is large).

Dewey and Bird regard the continents, less than 70 km thick, as superficial passengers on the plates (they assume the lithosphere may go down as far as 150 km) when the ocean floor spreads from such a ridge. The continents are prevented from sinking by their lightness (McKenzie, 1969). All tectonics will thus be governed by the interplay of the two crusts, oceanic and continental. When India, for example, borne by a plate which was going down under the Himalayan arc, came in contact with the Asian continent, the downward motion stopped and the northward expansion had to create a new trench elsewhere. According to Sykes (1970), this trench is in the process of formation along a diffuse line of epicenters running from the tip of India to Australia.

The problem of arcuate structures is taken up in discussions of the highly special case of the Marianas arc. The Marianas appear as an outlier of the Ryukyu arc, which according to Dewey and Bird, it would tend to approach in moving to the southwest as a result of sea floor spreading. Farther south the slab corresponding to the Manila trench is going down in the opposite sense; oceanic lithosphere between the northern Philippine arc and the continent tends to disappear. These appear to be two aspects of a general tendency of arcs to accumulate near continental margins.

Japan, where the volume and petrography of the various lavas have been fully studied, is taken next as a typical island arc example. As Kuno had done earlier, Dewey and Bird endeavor to account for the observations by partial melting in the descending slab. Also, between the 'volcanic front' forming the eastern boundary of the zone of active volcanoes and the trench 130 km farther east lies a prism of sediments (flysch) deriving from the volcanic arc and becoming thicker towards the trench. Finally, in the trench itself, only a part of the sediments descends. The remainder is scraped off the plate and plastered onto the inner trench wall, which is always steeper than the outer, then partially metamorphosed into blueschist. (Actually two blueschist zones are found in the Japanese arc, one Triassic, the other Cretaceous, which Miyashiro associates with two old trenches; Dewey and Bird see this as a final stage in the coming together of two distinct arcs.)

The growth of a single arc is supposed to begin with the development of a trench and the simultaneous beginning of plate descent. The inner wall of the trench, driven against the opposing plate by the downgoing slab, is formed into overlapping wedges, and constitutes a rise like that of Macquarie Island. Material returned to the trench by gravity slides contribute to the formation of blueschist. As the plate penetrates more deeply, basalts make their appearance to build up the main structure, the source of the flysch wedge. Next come the calc-alkaline magmas which become fractionated as they rise. Consequently a complex interplay of erosion, mixing and burial leads to the present situation.

Overriding of the arc by the inner wall of the trench is the essential feature of the mechanism of Dewey and Bird. Coleman (1971) emphasizes the parallel between the metamorphic blueschist zones and such peridotite massifs as those of Cyprus or New Caledonia, and he assumes that both are due to this overriding. All peridotite massifs of the Alpine type are supposed to originate by a similar process; this idea is sure to lead to discussion.

Oxburgh and Turcotte (1971) for their part attempt a quantitative explanation of the origin of the two metamorphic zones distinguished by Miyashiro. In their view the plastered sediments are formed into overriding wedges or downgoing wedges; in the region near the trench the latter at least are subjected to high pressure while remaining at low temperature; their metamorphism results in glaucophane blueschists returned to the surface by the rapid vertical movements accompanying each halt in the slab's descent. Inside the arc in a region heated by intrusions, mainly granitic, an alumino-silicate metamorphism would occur, requiring a high temperature but low pressure.

A quite different example of arcuate structure, the present day Andes, may serve as a model for all the Cordilleras, characterized according to Dewey and Bird by the underthrusting of a continental margin of the Atlantic type by the lithosphere. Actually they were guided principally by the structure of the Appalachian-Caledonian orogenic ensemble as reconstructed by Bullard *et al.* (1965). They first study the Atlantic margin of North America, assumed to have developed at the end of the Triassic with the separation from Africa. A prism of sediments up to 4 km thick in places is found under the coastal plain and the continental shelf. The outer edge of

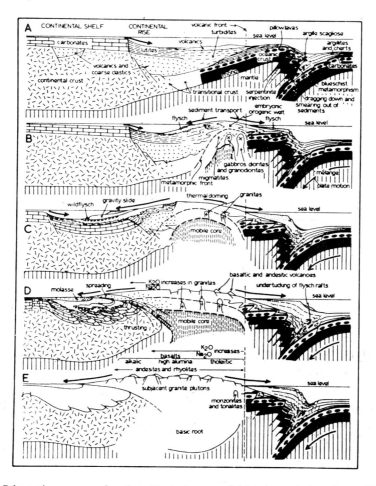

Fig. 119. Schematic sequence of sections illustrating a model for the evolution of a cordilleran-type mountain belt developed by the underthrusting of a continent by an oceanic plate. (Dewey and Bird, 1970; courtesy *Am. Geophys. Union.*)

the shelf is often marked by a buried ridge, supposed to be a reef. On the upper precontinental rise gravity slides occur forming fans at the foot of the rise and turbidites
farther down. Some fine sediments are transported from north to south on it by currents.
As such sources are inadequate to account for over 6 km of sediments in some troughs,
Dewey and Bird think these sediments are a mixture of volcanic and clastic materials
produced by erosion of the plateaus enclosing the Atlantic rift in the early stages of
separation. Subsidence during transport to the flanks of the ridge would bury these
coarse sediments. During this same early period of separation, when the seas were
shallow and the currents very weak, the salt was deposited which is now found in
domes all around the Atlantic rim: Labrador, Newfoundland, Mauretania, Morocco,
Portugal, Spain and Ireland (Pautot *et al.*, 1970), the Mediterranean (Montadert
et al., 1970). That the diapirs earlier discovered in the Gulf of Mexico are indeed
salt domes has been proved by JOIDES drilling.

Let us return to the Cordilleras (Figure 119). As with the island arcs, wedges of
oceanic crust and upper mantle are observed, pushed towards the precontinental rise,
with thickening of the flysch in the direction of the trench, and blueschist melanges.
When the sinking plate exceeds a depth of about 100 km, submarine volcanoes begin
to erupt behind the volcanic front. But a new essential feature occurs in eventual
ascent of a dome formed by thermal expansion, with a mobile core of melted gabbros
and granodiorites. This core metamorphoses the sediments, particularly the residues
from the initial rupture, and is then eroded itself both towards the ocean and towards

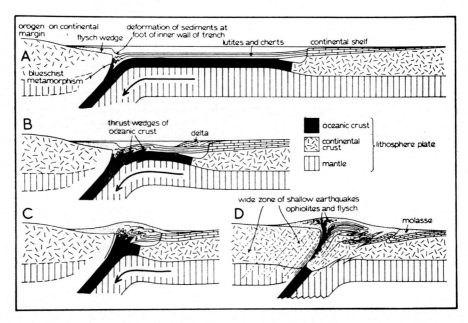

Fig. 120.   Schematic sequence of sections illustrating the collision of two continents. (Dewey and
Bird, 1970; courtesy *Am. Geophys. Union.*)

the subsiding and retreating continental margin. Thus some overriding in the direction of the continent is observed. The final stage includes the emplacement of granitic plutons, but the root of the range is essentially basic.

In this fashion the classical features of orogenesis are reproduced, including even the miogeosynclinal-eugeosynclinal pair (sedimentary, corresponding to the continental shelf; volcanic and clastic, corresponding to the region between shelf and trench).

Other attempts (as by Hsü in 1971) to account for the complexity of geosynclinal rocks without appealing to the action of an uplifted dome appear less convincing. This still qualitative dome hypothesis has been related to the older ideas of Wegmann (Griffin, 1970). Earlier, Elsasser (1970a) had suggested that the effect of foldings might be simulated by the accretion of materials under the crust.

This convective orogeny contrasts with the last important case, that of the collision of two continents. Dewey and Bird suppose one to have a continental margin of the Atlantic type, the other a trench with descending lithosphere (Figure 120). They emphasize the fact that the Mediterranean case is more complicated, involving ancient arcs and probably microcontinents. The basement of the overridden continent, unable to sink, is broken off in sheets which form the core of nappes, so that the root of the mountain belt is granitic in nature. Sinking lithospheric residue may be associated with local clusters of seismic foci (McKenzie, 1969). The shape of the opposing continents governs the extent and the intensity of the deformation and hence the solidity of the resulting suture.

11. Application of Plate Tectonics to Ancient Continental Movements

Plate tectonics is helpful in understanding continental displacements, even before the Jurassic, when the ocean crust can no longer furnish direct evidence of them. We shall give examples showing how geologic or paleomagnetic considerations may be complemented.

We begin with the breakup of the primitive continents. We may presume that a period of lithospheric extension and necking prepared the way for it, but it also may have come about abruptly. Experiments and calculations by Ramberg (1968) showed in fact that in a viscous medium the rate of rise of matter from a light layer depends strongly on the layer's thickness. It is thus possible to envisage the gradual build-up of an unstable asthenosphere followed by an outburst of diapiric activity, facilitated by the extremely rapid variation of viscosity with temperature. Another hypothesis may be mentioned as a curiosity, that of Gough (1970), accounting for the break-up of Gondwana in the Carboniferous by the tensions resulting from its icecap.

After breakup the lithosphere on either side is thick (Le Pichon and Fox, 1971; Le Pichon and Hayes, 1971); it thins later in the vicinity of the crest as the mid-ocean ridge becomes fully developed. A thick lithosphere constrains the two new continents to follow their transverse fractures closely, particularly when these are long. Thus the pole of the rotation which would bring the continents back into contact is stable, the more so if the pole is distant. If the pole is close by, the two edges of the lithosphere form a fairly wide V, facilitating the rise of deep matter into the open end. The V will thus tend to reclose and the pole will move off rapidly. It is considerations of this nature which aid in the reconstruction of ancient lithospheric plates.

To demonstrate the value of the evidence furnished by the fractures, we take up the best studied of all examples, the opening of the Atlantic Ocean. The opening of the North Atlantic and the related opening of the Labrador Sea have been investigated by the group of the Centre Océanologique de Bretagne. The great Gibbs fracture (Olivet *et al.*, 1970) on which three segments can be distinguished, and another transform fault which they named Farewell, were used by Le Pichon *et al.* (1971) to trace the history of the Labrador Sea. As to the North Atlantic, Le Pichon and Fox (1971) follow Heezen in utilizing matches between basement fractures which extend to the opposing continental margins. In the reconstruction of Bullard *et al.* (1965), these fractures do not coincide perfectly; the fit appears better in that of Dietz and Holden (1970). By starting from the best obtainable reconstruction and taking into account not only these transverse fractures but also the great longitudinal features to be seen in the Atlantic, particularly the boundaries between magnetically quiet and disturbed regions (Chapter II), Le Pichon and Fox end up with a model whose earlier stages only we recount. The opening began 180 m.y. ago in the North Atlantic according to JOIDES results, compared with 140 m.y. in the South Atlantic. The shallow initial opening corresponded to the magnetically quiet zone. Towards 76 m.y. ago (anomaly 32) there was already a sea between Rockall Bank and Europe. By this time the African and American continents were sufficiently separated to be able to rotate about a distant pole. Two-thirds of the Labrador Sea had been created by about −60 m.y.; around −49 m.y. (anomaly 20) it stopped growing while the ridge in the Norwegian Sea became active.

A similar undertaking for the South Atlantic has been carried out by Le Pichon and Hayes (1971); their task was made easier by the fit of Bullard *et al.* (1965), undisputed in this area. It is true that the fit leaves no place for the Caribbean, supposed to have been created in the Jurassic; at that time (Le Pichon and Fox, 1971) the Greater Antilles and Yucatan may have come into position from the west.

Le Pichon and Hayes (1971) also utilize opposite segments of the fracture zones, which take the form of trenches or rises, at times buried under the lateral basins where their presence can be detected by seismic reflection (Fail *et al.*, 1970). The Walvis and Rio Grande rises and the Falklands fracture zone are of this type, but not the volcanic belt stemming from Mount Cameroun. The initial spreading rate, 140 m.y. ago, assumed constant with time, varied from 0.9 cm/y. to 6.2 cm/y. from north to south, while it has remained fairly uniform since −80 m.y.

Although fracture zones are of great use, the only quantitative data for reconstructions depend on paleomagnetism, unfortunately restricted to the continents and whose application raises some delicate questions. Among these are: invariability of the plates; attachment of intercontinental ranges to one or the other of adjoining plates; uncertainty as to whether there is an absolute movement of the geographic pole with respect to the Earth, and if so, how to determine what it is with the plates drifting about, and so on.

Even the most complete paleomagnetic data leave some indeterminacy in the solution, but they are often imprecise and poorly dated, with the result that their direct employment still leaves room for hesitation in the reconstruction being sought. Francheteau and Sclater (1969), then Francheteau (1970) have developed a method which consists in extrapolating relative plate rotations from a known situation, either the present situation, or the one in which the continents under study were in contact. In so doing the constancy of certain instantaneous rotations in magnitude and direction are assumed constant over rather long periods, up to 100 m.y., for example. The postulate can be verified in the case of India, for which the geographic pole deduced from its rock magnetism describes a circle from the Jurassic onwards, and which has therefore pivoted about a point fixed within itself. Nevertheless the hypothesis remains arbitrary in some cases and cannot be generally true; more than two relative rotations among three plates cannot be strictly maintained (McKenzie and Morgan, 1969).

The situation at a particular epoch is recovered by simple successive rotations of the various continents with respect to one of their number. Even this step requires complicated calculations, as finite rotations are not vectorially additive. Paleo-magnetism is only invoked afterwards; the paleomagnetic poles for each continent at the epoch should coincide approximately in a point, the geographic pole. An arbitrary rotation about this pole is possible.

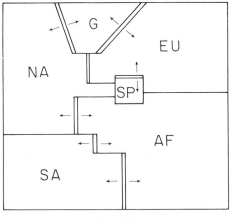

Fig. 121.

The opening of the Atlantic has also been considered by Francheteau. His model involves 6 plates (Figure 121), but neglects the Rockall bank. The present pole of the relative displacement between South America and Africa is not the pole required for complete closure of the ocean; for this Francheteau adopts the pole of Bullard *et al.* (1965), but the present pole may be retained back to −65 m.y., after which a second suffices for closure. The displacements Europe-Greenland, Greenland-North

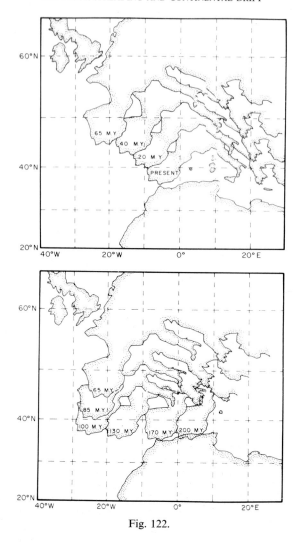

Fig. 122.

America and Spain-Europe are achieved with a single rotation, but the other relative movements are consequently complex.

The nature of the results so obtained may be seen in Figure 122 showing the movement of Europe with respect to Africa in 200 m.y. The opening of the Bay of Biscay is supposed to correspond to a rotation of only 22° between −100 and −65 m.y. (Further discussion of this problem would lead to the citation of a dozen recent papers!) Unfortunately North America cannot be shown on the map; it came into contact with Africa between −200 and −130 m.y.

Lastly, to give an idea of the precision (or otherwise) of present paleomagnetic results, Figure 123 shows the position of magnetic poles in the Upper Carboniferous for the continents as reassembled by Bullard *et al.*

In a thorough discussion of the results for the Lower Paleozoic grouping, Franche-

UPPER CARBONIFEROUS

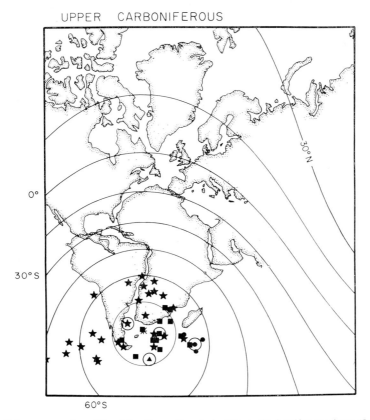

Fig. 123. Palaeogeography of the Atlantic continents in Upper Carboniferous times (fit of Bullard *et al.*). North America is kept fixed. The palaeolatitudes inferred from the palaeomagnetic poles from the four continents are drawn every 15°. The mean magnetic south pole (asterisk) and the south poles from North America (squares), Europe (stars), Africa (triangles) and South America (circles) appear on the map. The mean pole for each continent is surrounded by a circle. Mercator projection.
(Francheteau, 1970; courtesy J. Francheteau.)

teau points out the possibility of a slight improvement in the agreement among the corresponding poles by modifying Bullard's fit slightly. His main point however is the need to distinguish, within orogenies, the stages of the ancient plate boundaries. Within the Appalachian orogeny in particular, Dewey and Bird had first described a zone A as an Atlantic-type margin (ocean expanding) for the period between the end of the Precambrian and the beginning of the Ordovician, and then as a Pacific-type margin (with the ocean reclosing). They further distinguish southeast of zone A, a zone B of marine sediments and volcanics formed from the Cambrian to the Lower Devonian, and then a zone C, stable on the whole since the end of the Precambrian. Dewey found the corresponding zones in the Caledonian part of the same orogeny. The relative displacements of the three zones have been considerable, particularly in the Devonian, and there is no reason why they should give the same pole for this epoch as for previous ones. In Bullard's reconstruction the Cambrian poles for zone C of the two orogenies then adjacent are closer to each other than those of the Appalachian zones A and C.

Distinctions of this sort made by Francheteau within the tectonic regions of Eurasia and North Africa much reduce the dispersion in the poles for the Upper Paleozoic. We may hope then, that an improved choice of rock samples for study will lead to progress in the paleomagnetic reconstruction of the continents. For the present the

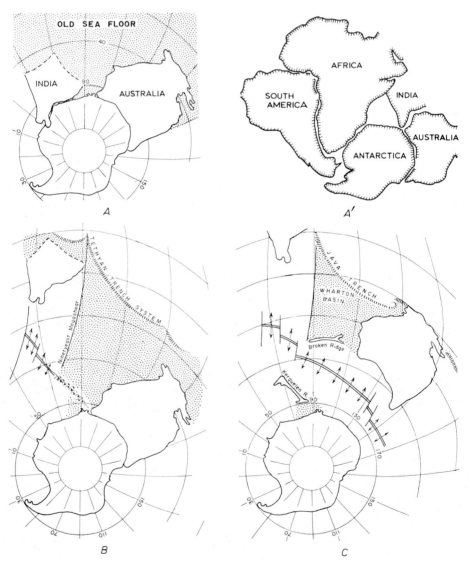

Fig. 124. A, B, C, time sequence diagram showing the breakup and drift dispersion of India and Australia from Antarctica and consequent preservation of old crust in the Wharton Basin (Dietz and Holden, 1970). A′, another possible configuration of Gondwanaland. Continental outlines at 500 fathom contour from Smith and Hallan. (Veevers *et al.*, 1971; courtesy *Nature*.)

uncertainties remain large, as will be seen in a final example, still dealing with Gond-wanaland.

Dietz and Holden (1971) seek to reason out a possible location of Paleozoic ocean crust, a problem which may have applications to mineral (in the basement) or petro-leum exploration (in sediments), in addition to its scientific interest. They conclude that the only hope is in the Wharton Basin west of Australia. Figure 124 shows the evolution of the Indian Ocean as it would follow from the generally accepted positions of India and Australia relative to Antarctica. But another possibility has been placed in the figure, a reconstruction for which Veevers *et al.* (1971) bring forward strati-graphic and tectonic arguments. In this solution the east side of India touches Australia instead of Antarctica, and a triangular sea is left in the center of Gondwana where Madagascar and the microcontinents of the Indian Ocean may be accommo-dated.

The only assured position seems to be that of Australia relative to Antarctica. Francheteau (1970) points out that a good fit between two continents depends on the rapidity of the transition from continental shelf to ocean. For South America and Africa the transition takes place over less than 100 km, while it often requires some-thing like 200 km for the North Atlantic margins; the south coast of Australia is an intermediate case.

REFERENCES

In order to keep the References relatively short, references are not necessarily to the basic papers in each field, but rather to *recent* articles providing access to all the others.

In addition to papers cited, the reader will find many interesting articles in the reference P. J. Hart (ed.), 1969.

Some abbreviations, which are not widely known:
J.G.R. = *Journal of Geophysical Research*
A.G.U. = *American Geophysical Union*
C.R. = *Comptes Rendus de l'Académie des Sciences.*
Ph.Tr. = *Philosophical Transaction of the Royal Society, A.*

Allan, D. W., Thompson, W. B. and Weiss, N. O.: 1967, 'Convection in the Earth's mantle', in *Mantles of the Earth and Terrestrial Planets* (ed. by S. K. Runcorn), Interscience Publishers, p. 507–512.

Allen, C. R.: 1965, 'Transcurrent Faults in Continental Areas', 82–89 in *A Symposium on Continental Drift, Ph. Tr.*, 1088.

Andel T. H. van and Heath, G. Ross: 1970, 'Tectonics of the Mid-Atlantic Ridge, 6–8° South Latitude', *Marine Geophys. Res.* 1, 5–36.

Anderson, D. L.: 1967, 'Latest Information from Seismic Observations', in *The Earth's Mantle* (ed. by T. F. Gaskell), Academic Press, p. 335–420.

Anonymous: 1964, 'Geological Society Phanerozoic Time-Scale', *Quart. J. Geol. Soc.*, London, 120 S, 260–262.

Artemjev, M. E. and Artyushkov, E. V.: 1971, 'Structure and Isostasy of the Baïkal Rift and the Mechanism of Rifting', *J.G.R.* **76**, 1197–1211.

Atwater, T. M. and Mudie, J. D.: 1968, 'Block Faulting on the Gorda Rise', *Science*, **159**, 729–731.

Bada, J. L., Luyendijk, B. P. and Maynard, J. B.: 1970, 'Marine Sediments: Dating by the Racemization of Amino-Acids', *Science* **170**, 730–732.

Baranzagi, M. and Isacks, B.: 1970, 'Lateral Variations of Seismic Wave Attenuation in the Upper Mantle Above the Inclined Earthquake Zone of the Tonga Island Arc', *EOS* (Trans. A.G.U.) **51**, 780.

Birch, F.: 1967, Low Values Oceanic Heat Flow, *J.G.R.* **72**, 2261–2262; X. Le Pichon, M. Langseth, 'Comments on Paper by Francis Birch . . .', *ibid.*, 6377–6378.

Björnsson, S. (ed.): 1967, *Iceland and Mid-Ocean Ridges*, Report of a Symposium, Visingdafélag Islendinga *(Soc. Sci. Islandica)*, Reykjavik, 209 p.

Boer, J. de, Schilling, J. G. and Krause, D. C.: 1969, 'Magnetic Polarity of Pillow Basalts from Reykjanes Ridge', *Science* **166**, 996–998.

Boldizsar, T.: 1968, 'Geothermal Data from the Vienna Basin, *J.G.R.* **73**, 613–618.

Bonhommet, N. and Babkine, J.: 1967, 'Sur la présence d'aimantations inversées dans la chaîne des Puys', *C.R.* **264**, 92–94.

Bosshard, E. and MacFarlane, D. J.: 1970, 'Crustal Structure of the Western Canary Islands from Seismic Refraction and Gravity Data', *J.G.R.* **75**, 4901–4918.

Bott, M. H. P.: 1965, 'Formation of Oceanic Ridges', *Nature* **207**, 840–843.

Bullard, E., Everett, J. E. and Smith, A. G.: 1965, 'The Fit of the Continents Around the Atlantic', 41–51 in *A Symposium on Continental Drift. Ph. Tr.*, 1088.

Burckle, L. H., Ewing, J., Saito, T. and Leyden, R.: 1967, 'Tertiary Sediment from the East Pacific Rise', *Science* **157**, 537–540.

Cann, J. R.: 1968, 'Geological Processes at Mid-Ocean Ridge Crests', *Geophys. J.* **15**, 331–341.

Christoffel, D. A. and Ross, D. I.: 1970, 'A Fracture Zone in the South West Pacific Basin South of

New Zealand and Its Implications for Sea Floor Spreading', *Earth Planetary Sci. Letters* **8**, 125–130.

Clark, S. P. and Ringwood, A. E.: 1967, 'Density, Strength, and Constitution of the Mantle', in *The Earth's Mantle* (ed. by T. F. Gaskell), Academic Press, p. 111–124.

Coleman, R. G.: 1971, 'Plate Tectonics Emplacement of Upper Mantle Peridotites along Continental Edges', *J.G.R.* **76**, 1212–1222.

Cook, K. L.: 1962, 'The Problem of the Mantle-Crust Mix: Lateral Inhomogeneity in the Uppermost Part of the Earth's Mantle', *Advan. Geophys.* **9**, 295–360.

Coulomb, J. and Jobert, G.: 1963, *Physical Constitution of the Earth*, Oliver & Boyd, 328 p.

Coulomb, J. and Jobert, G.: 1967, L'énergie libérée dans les séismes et la théorie d'Orowan, *Bull. Soc. Roy. Sci. Liège* **36** 32–37.

Coulomb, J. and Loisel, J.: 1940, *La physique des nuages*, Albin Michel, 286 p.

Cox, A. and Dalrymple, G. B.: 1967, 'Geomagnetic Polarity Epochs: Nunivak Island, Alaska', *Earth Planetary Sci. Letters* **3**, 173–177.

Cox, A., Doell, R. R. and Dalrymple, G. B.: 1964, 'Reversals of the Earth's Magnetic Field', *Science* **144**, 1537–1543.

Creer, K. M.: 1965, Palaeomagnetic Data from Gondwanic Continents, 27–40 in *A Symposium on Continental Drift, Ph. Tr.*, 1088.

Dagley, P., Wilson, R. L., Ade-Hall, J. M., Walker, G. P. L., Haggerty, S. E., Sigurgeirsson, T., Watkins, N. D., Smith, P. J., Edwards, J. and Grasty, R. L.: 1967, 'Geomagnetic Polarity Zones for Icelandic Lavas', *Science* **216**, 25–29.

Dehlinger, P., Couch, R. W. and Gemperle, M.: 1967, 'Gravity and Structure of the Eastern Part of the Mendocino Escarpment, *J.G.R.* **72** 1233–1247.

Denham, C. R. and Cox, A.: 1970, 'Palaeomagnetic Evidence that the Laschamp Polarity Event Did Not Occur between 30000 and 12000 Years Ago', *EOS* **51**, 745.

Dewey, J. F. and Bird, J. M.: 1970, 'Mountain Belts and the New Global Tectonics', *J.G.R.* **75**, 2625–2647.

Dickson, G. O., Pitman III, W. C. and Heirtzler, J. R.: 1968, 'Magnetic Anomalies in the South Atlantic and Ocean Floor Spreading', *J.G.R.* **73**, 2087–2100.

Dietz, R. S.: 1966, 'Passive Continents, Spreading Sea Floors, and Collapsing Continental Rises', *Am. J. Sci.* **264**, 177–193; Discussion by G. M. Young, *ibid.*, 1967, **265**, 225–230; A reply by R. S. Dietz, *ibid.*, **265**, 231–237.

Dietz, R. S. and Holden, J. C.: 1970, 'Reconstruction of Pangaea: Breakup and Dispersion of Continents, Permian to Present', *J.G.R.* **75**, 4939–4956.

Dietz, R. S. and Holden, J. C.: 1971, 'Pre-Mesozoic Oceanic Crust in the Eastern Indian Ocean (Wharton Basin)?', *Nature* **229**, 309–312.

Dietz, R. S. and Sproll, W. P.: 1970, 'East Canary Islands as a Microcontinent within the Africa-North America Continental Drift Fit', *Nature* **226**, 1043–1045.

Dubois, J.: 1966, 'Temps de propagation des ondes *P* à des distances épicentrales de 30 à 90°. Région du Sud-Ouest Pacifique', *Ann. Geophys.* **22**, 642–645.

Dubois, J.: 1969, 'Contribution à l'étude structurale du sud-ouest Pacifique d'après les ondes sismiques observées en Nouvelle-Calédonie et aux Nouvelles-Hébrides', Thèse, Paris.

Dymond, J. and Windom, H. L.: 1968, 'Cretaceous K-Ar Ages from Pacific Ocean Sea-Mounts', *Earth Planetary Sci. Letters* **4**, 47–52.

Elder, J.: 1967, 'Convective Self-Propulsion of Continents', *Nature* **214**, 657–660, 750.

Elsasser, W. M.: 1968, *Convection and Stress Propagation in the Upper Mantle*, Newcastle symposium, New York, Wiley (in press).

Elsasser, W. M.: 1970a, 'The So-Called Folded Mountains', *J.G.R.* **75**, 1615–1618.

Elsasser, W. M.: 1970b, 'Non-Uniformity of Crustal Growth', *EOS (Trans. A.G.U.)* **51**, 823.

Elsasser, W. M.: 1971, 'Sea Floor Spreading as Thermal Convection', *J.G.R.* **76**, 1101–1112.

Emilia, D. A. and Heinrichs, D. F.: 1969, 'Ocean Floor Spreading: Olduvai and Gilsa Events in the Matuyama Epoch', *Science* **166**, 1267–1269.

Engel, C. G. and Fisher, R. L.: 1969, 'Lherzolite, Anorthosite, Gabbro, and Basalt Dredged from the Mid-Indian Ocean Ridge', *Science* **166**, 1136–1141.

Epp, D., Grim, P. J. and Langseth Jr., M. G.: 1970, 'Heat Flow in the Caribbean and Gulf of Mexico', *J.G.R.* **75**, 5655–5669.

Evans, A. L.: 1970, 'Geomagnetic Polarity Reversals in a Late Tertiary Lava Sequence from the Akaroa Volcanoes, New Zealand', *Geophys. J.* **21**, 163–183.

Ewing, J. and Ewing, M.: 1967, 'Sediment Distribution on the Mid-Ocean Ridges with Respect to Spreading of the Sea Floor', *Science* **156**, 1590–1592.

Ewing, J., Windisch, C. and Ewing, M: 1970, 'Correlation of Horizon A with JOIDES Bore-Hole Results', *J.G.R.* **75**, 5645–5653.

Ewing, J., Ewing, M., Aiken, T. and Ludwig, W.: 1968, 'North Pacific Sediment Layers Measured by Seismic Profiling', delivered at 11th Pacific Science Congress, Tokyo, August–September 1966, *A.G.U. Geophysical Monograph* **12**, 147–173.

Fail, J. P., Montadert, L., Delteil, J. R., Valéry, P., Patriat, P. and Schlich, R.: 1970, 'Prolongation des zones de fracture de l'Océan Atlantique dans le Golfe de Guinée', *Earth Planetary Sci. Letters* **7**, 413–419.

Falcon, N. L.: 1967, 'Equal Areas of Gondwana and Laurasia', *Nature* **213**, 580–581.

Falcon, N. L., Gass, I. G., Girdler, R. W. and Laughton, A. S. (eds.): 1970, 'A Discussion on the Structure and Evolution of the Red Sea and the Nature of the Red Sea, Gulf of Aden and Ethiopia Rift Junction', *Ph. Tr.* **267**, 1–417.

Fischer, A. G., Heezen, B. C., Boyce, R. E., Bukry, D., Douglas, R. G., Garrison, R. E., Kling, S. A., Krasheninnikov, V., Lisitzin, A. P. and Pimm, A. C.: 1970, 'Geological History of the Western North Pacific', *Science* **168**, 1210–1214.

Fisher, R. L. and Engel, C. G.: 1968, 'Dunite Dredged from the Nearshore Flank of Tonga Trench on Expedition Nova 1967', *Trans. A.G.U.* **49**, 217–218.

Fitch, T. J.: 1970, 'Earthquake Mechanisms in the Himalayan, Burmese, and Andaman Regions and Continental Tectonics in Central Asia', *J.G.R.* **75**, 2699–2709.

Fitch, T. J. and Molnar, P.: 1970, 'Focal Mechanisms along Inclined Earthquake Zones in the Indonesia-Philippine Region', *J.G.R.* **75**, 1431–1444.

Fox, P. J.: 1967, 'Annotated Bibliography on the World Rift System', World Data Center A; Upper Mantle Project Report No. 14.

Fox, P. J. and Heezen, B. C.: 1965, 'Sands of the Mid-Atlantic Ridge', *Science* **149**, 1367–1370.

Francheteau, J.: 1970, Ph.D. Thesis, University of California, San Diego.

Francheteau, J. and Sclater, J. G.: 1969, 'Palaeomagnetism of the Southern Continents and Plate Tectonics', *Earth Planetary Sci. Letters* **6**, 93–106.

Francheteau, J., Sclater, J. G. and Menard, H. W.: 1970, 'Pattern of Relative Motion from Fracture Zone and Spreading Rate Data in the North-Eastern Pacific', *Nature* **226**, 746–748.

Gardner, J. V.: 1970, 'Submarine Geology of the Western Coral Sea', *Bull. Geol. Soc. Am.* **81**, 2599–2614.

Girdler, R. W.: 1968, 'Drifting and Rifting of Africa', *Nature* **217**, 1102–1106.

Glass, B.: 1967, 'Tektites and Geomagnetic Reversals', *Nature* **214**, 372–374.

Glass, B. P. and Heezen, B. C.: 1967, 'Tektites and Geomagnetic Reversals', *Sci. Am.* **217**, 32–38.

Godby, E. A., Baker, R. C., Bower, M. E. and Hood, P. J.: 1966, 'Aeromagnetic Reconnaissance of the Labrador Sea', *J.G.R.* **71**, 511–517.

Goguel, J.: 1965, 'Tectonics and Continental Drift', 194–198 in *A Symposium on Continental Drift*, *Ph. Tr.*, 1088.

Gough, D. I.: 1970, 'Did an Ice-Cap Break Gondwanaland?' *J.G.R.* **75**, 4475–4477.

Griffin, V. S.: 1970, 'Relevancy of the Dewey-Bird Hypothesis of Cordilleran-Type Mountain Belts and the Wegmann Stockwork Concept', *J.G.R.* **75**, 7504–7507.

Grim, P. and Erickson, B. H.: 1969, 'Fracture Zones and Magnetic Anomalies South of the Aleutian Trench', *J.G.R.* **74**, 1488–1494.

Hamilton, E. L. and Menard, H. W.: 1968, 'Undistorted Turbidites on the Juan de Fuca Ridge', *Trans. A.G.U.* **49**, 208.

Hammond, A. L.: 1970, 'Deep Sea Drilling, a Giant Step in Geological Research', *Science* **170**, 520–521.

Harrison, C. G. A. and Mudie, J. D.: 1967, 'Sediment Thickness and Sea-Floor Spreading in the Pacific', *Trans. A.G.U.* **48**, 133.

Hart, P. J. (ed.): 1969, 'The Earth's Crust and Upper Mantle', *Geophysical Monograph* **13**, 735 p., *A.G.U.*

Hatherton, T.: 1969, 'Similarity of Gravity Anomaly Patterns in Asymmetric Active Regions', *Nature* **224**, 357–358.

Hayes, D. E. and Ludwig, W. J.: 1967, 'The Manila Trench and West Luzon Trough. II: Gravity and Magnetic Measurements, *Deep-Sea Res.* **14**, 545–560.

Hayes, D. E. and Heirtzler, J. R.: 1968, 'Magnetic Anomalies and Their Relation to the Aleutian Island Arc', *Trans. A.G.U.* **49**, 207–208.

Heezen, B. C., Tharp, M. and Ewing, M.: 1959, 'The Floors of the Oceans. 1: The North Atlantic', *Geol. Soc. Am.*, Spec. Pap. 65, 122 p.

Heezen, B. C., Gray, C., Segre, A. G. and Zarudski, E. F. K.: 1971, 'Evidence of Foundered Continental Crust beneath the Central Tyrrhenian Sea', *Nature* **229**, 327–329.

Heirtzler, J. R.: 1965, 'Marine Geomagnetic Anomalies', *J. Geomagn. Geoelec.* **17**, 227–236.

Heirtzler, J. R.: 1970, 'The Paleomagnetic Field as Inferred from Marine Studies', *J. Geomagn. Geoelec.* **22**, 197–211,

Heirtzler, J. R. and Hayes, D. E.: 1967, 'Magnetic Boundaries in the North Atlantic Ocean', *Science* **157**, 185–187.

Heirtzler, J. R. and Pichon, X. Le: 1965, 'Crustal Structure of the Mid-Ocean Ridges. 3: Magnetic Anomalies over the Mid-Atlantic Ridge', *J.G.R.* **70**, 4013–4033.

Heirtzler, J. R., Pichon, X. Le and Baron, J. G.: 1966, 'Magnetic Anomalies over the Reykjanes Ridge', *Deep-Sea Res.* **13**, 427–443.

Heirtzler, J. R., Dickson, G. O., Herron, E. M., Pitman III, W. C. and Pichon, X. Le: 1968, 'Marine Magnetic Anomalies, Geomagnetic Field Reversals, and Motions of the Ocean Floor and Continents', *J.G.R.* **73**, 2119–2136.

Herron, E. M. and Hayes, D. H.: 1969, 'A Geophysical Study of the Chile Ridge', *Earth Planetary Sci. Letters* **6**, 77–83.

Herron, E. M. and Heirtzler, J. R.: 1967, 'Sea-Floor Spreading near the Galapagos', *Science* **158**, 775–780.

Herzen, R. P. von, Simmons, G. and Folinsbee, A.: 1970, 'Heat Flow Between the Caribbean Sea and Mid-Atlantic Ridge', *J.G.R.* **75**, 1973–1984,

Hess, H. H.: 1962, 'History of Ocean Basins', in *Petrologic Studies, A Volume to Honor A. F. Buddington*, p. 599–620, Geol. Soc. Am.

Hess, H. H.: 1964, 'Seismic Anisotropy of the Uppermost Mantle under Oceans', *Nature* **203**, 629–631.

Hess, H. H.: 1965, 'Mid Oceanic Ridges and Tectonics of the Sea-Floor', in *Submarine Geology and Geophysics*, p. 317–333, Colston Paper 17, Butterworths.

Hurley, P. M. and Rand, J. R.: 1969, 'Pre-Drift Continental Nuclei', *Science* **164**, 1229–1242.

Hurley, P. M., Almeida, F. F. M. de, Melcher, G. C., Cordani, U. G., Rand, J. R., Kawashita, K., Vandoros, P., Pinson Jr, W. H. and Fairbairn, H. W.: 1967, 'Test of Continental Drift by Comparison of Radiometric Ages', *Science* **157**, 495–500.

Isacks, B. and Molnar, P.: 1970, 'Mantle Earthquake Mechanisms and the Sinking of the Lithosphere', *Nature* **223**, 1121–1124.

Isacks, B., Oliver, J. and Sykes, L. R.: 1968, 'Seismicity and the New Global Tectonics', *J.G.R.* **73**, 5855–5899.

Jacoby, W. R.: 1970, 'Instability in the Upper Mantle and Global Plate Movements', *J.G.R.* **75**, 5671–5680.

Johnson, C. L. and Pew, J. A.: 1968, 'Extension of the Mid-Labrador Sea Ridge', *Nature* **217**, 1033–1034.

Johnson, C. L. and Vogt, P. R.: 1968, 'Present and Relic Sea-Floor Rifts in the Arctic', *Trans. A.G.U.* **49**, 201.

Jones, J. G.: 1971, 'Aleutian Enigma: a Clue to Transformation in Time', *Nature* **229**, 400–403.

Kanamori, H. and Press, F.: 1970, 'How Thick is the Lithosphere?', *Nature* **226**, 330–331.

Karig, D. E.: 1970, 'Ridges and Basins of the Tonga-Kermadec Island Arc System', *J.G.R.* **75**, 239–254.

Kaula, W. M.: 1970, 'Earth's Gravity Field: Relation to Global Tectonics', *Science* **169**, 982–985.

Kaula, W. M.: 1971, 'Global Gravity and Tectonics', paper presented at the Francis Birch Symposium on the Nature of the Solid Earth, April 16–18, 1970, Harvard University, Cambridge, Mass., in press.

Kay, R., Hubbard, N. J. and Gast, P. W.: 1970, 'Chemical Characteristics and Origin of Oceanic Ridge Volcanic Rocks', *J.G.R.* **75**, 1585–1613.

Knopoff, L.: 1967, 'Thermal Convection in the Earth's Mantle', in *The Earth's Mantle* (ed. by T. F. Gaskell), Academic Press, p. 171–196.

Krause, D. C. and Watkins, N. D.: 1970, 'North Atlantic Crustal Genesis in the Vicinity of the Azores', *Geophys. J.* **19**, 261–283.

Ku, T. L., Broecker, W. S. and Opdyke, N.: 1968, 'Comparison of Sedimentation Rates Measured by Palaeomagnetic and the Ionium Methods of Age Determination', *Earth Planetary Sci. Letters* **4**, 1–16.

Kuno, H.: 1967, 'Volcanological and Petrological Evidence Regarding the Nature of the Upper Mantle', in *The Earth's Mantle* (ed. by T. F. Gaskell) Academic Press, p. 89–110.

Labrouste, Y. H., Baltenberger, P., Perrier, G. and Recq, M.: 1968, 'Courbes d'égale profondeur de la discontinuité de Mohorovicic dans le sud-est de la France', *C.R.* **266**, 663–665.

Langseth, M. G., Pichon, X. Le and Ewing, M.: 1966, 'Crustal Structure of the Mid-Ocean Ridges. 5: Heat Flow Through the Atlantic Ocean Floor and Convection Currents', *J.G.R.* **71**, 5321–5355.

Langseth Jr, M. G. and Taylor, P. T.: 1967, 'Recent Heat Flow Measurements in the Indian Ocean', *J.G.R.* **72**, 6249–6260.

Larson, R. L. and Spiess, F. N.: 1969, 'East Pacific Rise Crest: A Near Bottom Geophysical Profile', *Science* **163**, 68–71.

Lee, W. H. K. and Uyeda, S.: 1965, 'Review of Heat Flow Data', in *Terrestrial Heat, Geophysical Monograph* **8**, p. 87–190, *A.G.U.*

Lehmann, I.: 1967, 'Low-Velocity Layers', in *The Earth's Mantle* (ed. by T. F. Gaskell), Academic Press, p. 41–61.

Lister, C. R. B.: 1970, 'Heat Flow West of the Juan de Fuca Ridge', *J.G.R.* **75**, 2648–2654.

Lliboutry, L.: 'Introduction à la mécanique des plaques' (ed. by Mattauer and Allègre), Colloque Paris 1970, in press.

Lubimova, E. A.: 1967, 'Theory of Thermal State of the Earth's Mantle', in *The Earth's Mantle* (ed. by T. F. Gaskell), Academic Press, p. 231–323.

Ludwig, W. J., Hayes, D. E. and Ewing, J. I.: 1967, 'The Manila Trench and West Luzon Trough. I: Bathymetry and Sediment Distribution, *Deep-Sea Res.* **14**, 533–544.

Luyendyk, B. P.: 1969, 'Origin of Short-Wavelength Magnetic Lineations Observed near the Ocean Bottom', *J.G.R.* **74**, 4869–4881.

Luyendyk, B. P.: 1970, *EOS (Trans. A.G.U.)* **51**, 325.

Luyendyk, B. P. and Fischer, D. E.: 1969, 'Fission Track Age of Magnetic Anomaly 10: A New Point on the Sea-Floor Spreading Curve'. *Science* **164**, 1516–1517.

Malahoff, A.: 1970, 'Some Possible Mechanisms for Gravity and Thrust Faults under Oceanic Trenches', *J.G.R.* **75**, 1992–2001.

Malkus, W. V. R.: 1971, in *Mantle and Core in Planetary Physics, Enrico Fermi Summer Course* (ed. by M. Caputo and J. Coulomb), Academic Press, in press.

Matthews, D. H. and Bath, J.: 1967, 'Formation of Magnetic Anomaly Pattern of Mid-Atlantic Ridge', *Geophys. J.* **13**, 349–357.

Maxwell, A. E., Herzen, R. P. von, Hsü, K. J., Andrews, J. E., Saito, T., Percival Jr, S. F., Milow, E. D. and Boyce, R. E.: 1970, 'Deep Sea Drilling in the South Atlantic', *Science* **168**, 1047–1059.

McConnell Jr, R. K., McClaine, L. A., Lee, D. W., Aronson, J. R. and Allen, R. V.: 1967, 'A Model for Planetary Igneous Differentiation', *Rev. Geophys.* **5**, 121–172.

McKenzie, D. P.: 1967a, 'The Viscosity of the Mantle', *Geophys. J.* **14**, 297–305.

McKenzie, D. P.: 1967b, 'Some Remarks on Heat Flow and Gravity Anomalies', *J.G.R.* **72**, 6261–6273.

McKenzie, D. P.: 1969, 'Speculations on the Consequences and Causes of Plate Motions', *Geophys. J.* **18**, 1–32.

McKenzie, D. P.: 1970, 'Plate Tectonics', in *Birch Symp.*, in press.

McKenzie, D. P. and Parker, R. L.: 1967, 'The North Pacific: an Example of Tectonics on a Sphere', *Nature* **216**, 1276–1280.

McKenzie, D. P., Davies, D. and Molnar, P.: 1970, 'Plate Tectonics of the Red Sea and East Africa', *Nature* **226**, 243–248.

McKenzie, D. P. and Morgan, W. J.: 1969, 'Evolution of Triple Junctions', *Nature* **224**, 125–133.

Medaris Jr, L. G. and Dott Jr, R. H.: 1970, 'Mantle-Derived Peridotites in Southwestern Oregon: Relation to Plate Tectonics', *Science* **169**, 971–974.

Melson, W. G. and Thompson, G.: 1970, 'Layered Basic Complex in Oceanic Crust, Romanche Fracture, Equatorial Atlantic Ocean', *Science* **168**, 817–820.

Menard, H. W.: 1964, *Marine Geology of the Pacific*, McGraw-Hill, New York, 271 p.

Menard, H. W.: 1965a, 'The World-Wide Oceanic Rise-Ridge System', in *A Symposium on Continental Drift*, 109–122, *Ph. Tr.*, 1088.

Menard, H. W.: 1965b, 'Sea Floor Relief and Mantle Convection, *Phys. Chem. Earth*, **6**, 315–364.

Menard, H. W.: 1966, 'Fracture Zones and Offsets of the East Pacific Rise', *J.G.R.* **71**, 682–685.

Menard, H. W.: 1967a, 'Extension of Northeastern-Pacific Fracture Zones', *Science* **155**, 72–74.

Menard, H. W.: 1967b, 'Sea Floor Spreading, Topography, and the Second Layer', *Science* **257**, 923–924.

Menard, H. W.: 1969, 'Growth of Drifting Volcanoes', *J.G.R.* **74**, 4833–4837.

Menard, H. W., Winterer, E. L., Chase, T. E. and Smith, S. M.: 1968, 'Melanesian Sea-Floor Relief', *Trans. A.G.U.* **49**, 217.

Mercy, E. L. P.: 1967, 'Geochemistry of the Mantle', in *The Earth's Mantle* (ed. by T. F. Gaskell), Academic Press, p. 421–443

Milson, J. S.: 1970, 'Woodlark Basin, a Minor Center of Sea Floor Spreading in Melanesia', *J.G.R.* **75**, 7335–7339.

Mitronovas, W., Isacks, B. and Seeber, L.: 1969, 'Earthquake Locations and Seismic Wave Propagation in the Upper 250 km of the Tonga Island Arc', *Bull. Seism. Soc. Am.* **59**, 1115–1135.

Miyashiro, A., Shido, F. and Ewing, M.: 1970, 'Petrologic Models for the Mid-Atlantic Ridge', *Deep-Sea Res.* **17**, 109–123.

Mohammadioun, B.: 1966, *Structure du manteau et du noyau terrestres d'après les spectres d'énergie des ondes longitudinales*, thèse, Paris.

Molnar, P. and Oliver, J.: 1969, 'Lateral Variations of Attenuation in the Upper Mantle and Discontinuities in the Lithosphere', *J.G.R.* **74**, 2648–2682.

Montadert, L., Sancho, J., Fail, J. P., Debyser, J. and Winnock, E.: 1970, 'De l'âge tertiaire de la série salifère responsable des structures salifères en Méditerranée occidentale (Nord-Est des Baléares)', *C.R.* **271D**, 812–815.

Montigny, R., Javoy, M. and Allègre, C. J.: 1969, 'Le problème des andésites. Etude du volcanisme quaternaire du Costa Rica (Amérique Centrale) à l'aide des traceurs couplés $^{87}Sr/^{86}Sr$ et $^{18}O/^{16}O$', *Bull. Soc. Géol. France* (7), **11**, 794–799.

Morgan, W. J.: 1968, 'Rises, Trenches, Great Faults, and Crustal Blocks', *J.G.R.* **73**, 1959–1982.

Morgan, W. J.: 1970, 'Plate Motions and Deep Mantle Convection', *EOS (Trans. A.G.U.)* **51**, 822.

Morris, G. B., Raitt, R. W. and Shor Jr. G. G.: 1969, 'Velocity Anisotropy and Delay-Time Maps of the Mantle near Hawaii', *J.G.R.* **74**, 4300–4316.

Northrop, J., Morrison, M. F. and Duennebier, F. K.: 1970, 'Seismic Slip Rate on the Eastern Pacific Rise and Pacific Antarctic Ridge', *J.G.R.* **75**, 3285–3290.

Nur, A.: 1971, 'Viscous Phase in Rocks and the Low-Velocity Zone', *J.G.R.* **76**, 1270–1271.

Oliver, J. and Isacks, B.: 1967, 'Deep Earthquake Zones, Anomalous Structures in the Upper Mantle, and the Lithosphere', *J.G.R.* **72**, 4259–4275.

Olivet, J. L., Sichler, B., Thonon, P., Pichon, X. Le, Martinais, J. and Pautot, G.: 1970, 'La faille transformante Gibbs entre le Rift et la marge du Labrador', *C.R.* **271D**, 949–952.

Opdyke, N. D. and Wilson, K.: 1968, 'A Test of the Dipole Hypothesis', *Trans. A.G.U.* **49**, 125.

Opdyke, N. D., Glass, B., Hays, J. D. and Foster J.: 1966, 'Paleomagnetic Study of Antarctic Deep-Sea Cores', *Science* **154**, 349–357.

Orowan, E.: 1965, 'Convection in a Non-Newtonian Mantle, Continental Drift, and Mountain Building', in *A Symposium on Continental Drift*, 284–313, *Ph. Tr.*, 1088.

Orowan, E.: 1966, 'Age of the Ocean Floor', *Science* **154**, 413–416.

Oxburgh, E. R. and Turcotte, D. L.: 1968, 'Problem of High Heat Flow and Volcanism Associated with Zones of Descending Mantle Convective Flow', *Nature* **218**, 1041–1043.

Oxburgh, E. R. and Turcotte, D. L.: 1971, 'Origin of Paired Metamorphic Belts and Crustal Dilation in Island Arcs Regions', *J.G.R.* **76**, 1315–1327.

Pautot, G., Auzende, J. and Pichon, X. Le: 1970, 'Continuous Deep Sea Salt Layer along North Atlantic Margins Related to Early Phase of Rifting', *Nature* **227**, 351–354.

Peter, G., Elvers, D. and Yellin, M.: 1965, 'Geological Structure of the Aleutian Trench Southwest of Kodiak Island', *J.G.R.* **70**, 353–366.

Peychès, I. and Zortea, M.: 1971, 'Glass Tanks as Models for Convection in the Upper Mantle', *J.G.R.* **76**, 1416–1423.

Phillips, J. D.: 1967, 'Magnetic Anomalies over the Mid-Atlantic Ridge near 27°N', *Science* **157**, 920–923.

Pichon, X. Le: 1966, *Étude géophysique de la dorsale médio-atlantique*, thèse, Strasbourg, Lamont Contribution No. 922.

Pichon, X. Le: 1968, 'Sea-Floor Spreading and Continental Drift', *J.G.R.* **73**, 3661–3697.

Pichon, X. Le and Fox, J.: 1971, *Marginal Offsets, Fracture Zones, and the Early Opening of the North Atlantic*, in press.

Pichon, X. Le and Hayes, D. E.: 1971, *Marginal Offsets, Fracture Zones, and the Early Opening of the South Atlantic*, in press.

Pichon, X. Le and Heirtzler, J. R.: 1968, 'Magnetic Anomalies in the Indian Ocean and Sea-Floor Spreading', *J.G.R.* **73**, 2101–2117.

Pichon, X. Le and Langseth, M. G.: 1968, *Heat Flow from the Mid-Ocean Ridges and Sea-Floor Spreading*, preprint.

Pichon, X. Le, Hyndman, R. D. and Pautot, G.: 1971, *A Geophysical Study of the Opening of the Labrador Sea*, in press.

Pichon, X. Le, Houtz, R. E., Drake, C. L. and Nafe, J. E.: 1965, 'Crustal Structure of the Mid-Ocean Ridges. 1: Seismic Measurements', *J.G.R.* **70**, 319–339.

Pichon, X. Le, Cressard, A., Mascle, J., Pautot, G. and Sichler, B.: 1970, 'Structures sous-marines des bassins sédimentaires de Porcupine et de Rockall', *C.R.* **270D**, 2903–2906.

Pitman III, W. C. and Heirtzler, J. R.: 1966, 'Magnetic Anomalies over the Pacific-Antarctic Ridge', *Science* **154**, 1164–1171.

Pitman III, W. C., Herron, E. M. and Heirtzler, J. R.: 1968, 'Magnetic Anomalies in the Pacific and Sea-Floor Spreading', *J.G.R.* **73**, 2069–2085.

Price, A. T.: 1967, 'Magnetic Variations and Telluric Currents', in *The Earth's Mantle* (ed. by T. F. Gaskell), Academic Press, p. 125–170.

Ramberg, H.: 1968, 'Instability of Layered Systems in the Field of Gravity', *Phys. Earth Planetary Int.* **1**, 427–474.

Rea, D. K.: 1970, 'Changes in Structure and Trend of Fracture Zones North of the Hawaiian Ridge and Relation to Sea-Floor Spreading', *J.G.R.* **75**, 1421–1430.

Reid Jr, J. B. and Frey, F. A.: 1971, 'Rare Earth Distributions in Lherzolite and Garnet Pyroxenite Xenoliths and the Constitution of the Upper Mantle', *J.G.R.* **76**, 1184–1196.

Riedel, W. R.: 1967, 'Radiolarian Evidence Consistent with Spreading of the Pacific Floor', *Science* **157**, 540–542.

Rikitake, T.: 1966, *Electromagnetism and the Earth's Interior*, Elsevier, 308 p.

Rona, P. A., Brakl, J. and Heirtzler, J. R.: 1970, 'Magnetic Anomalies in the North-East Atlantic Between the Canary and Cape Verde Islands', *J.G.R.* **75**, 7412–7420.

Runcorn, S. K.: 1965, 'Palaeomagnetic Comparisons between Europe and North-America', in *A Symposium on Continental Drift*, 1–11 *Ph. Tr.*, 1088.

Runcorn, S. K. (ed.): 1967, *International Dictionary of Geophysics*, Pergamon Press, 2 vol., 1728 p.

Saito, T., Ewing, M. and Burckle, L. H.: 1966, 'Tertiary Sediment from the Mid-Atlantic Ridge', *Science* **151**, 1075–1079.

Scarfe, C. M. and Wyllie, P. J.: 1967, 'Serpentine Dehydration Curves and Their Bearing on Serpentinite Deformation in Orogenesis', *Nature* **215**, 945–946.

Schilling, J. G., Krause, D. C. and Moore, J. G.: 1968, 'Geological, Geochemical, and Magnetic Studies of the Reykjanes Ridge near 60°N', *Trans. A.G.U.* **49**, 201–203.

Schlich, R. and Patriat, P.: 1968, 'Interprétation possible de données géophysiques recueillies sur la dorsale médio-indienne entre 20° et 40° Sud', *C.R.* **266B**, 820–822.

Schlich, R. and Patriat, P.: 1971, 'Anomalies magnétiques de la branche Est de la dorsale médio-indienne entre les îles Amsterdam et Kerguelen', *C.R.* **272B**, 773–776.

Schneider, E. D. and Vogt, P. R.: 1968, 'Discontinuities in the History of Sea-Floor Spreading', *Nature* **217**, 1212–1222.

Scholl, D. W. and Huene, R. von: 1968, 'Spreading of the Ocean Floor: Undeformed Sediments in the Peru-Chile Trench', *Science* **159**, 689–871.

Schubert, G. and Turcotte, D. L.: 1971, 'Phase Changes and Mantle Convection', *J.G.R.* **76**, 1424–1432.

Sclater, J. G. and Francheteau, J.: 1970, 'The Implications of Terrestrial Heat Flow Observations on Current Tectonic and Geochemical Models of the Crust and Upper Mantle of the Earth', *Geophys. J.* **20**, 509–542.

Sclater, J. G. and Menard, H. W.: 1967, 'Topography and Heat Flow on the Fiji Plateau', *Nature* **216**, 991–993.

Serson, P. H., Hannaford, W. and Haines, G. V.: 1968, 'Magnetic Anomalies over Iceland', *Science* **162**, 355–357.

Shaw, H. R.: 1970, 'Earth Tides, Global Heat Flow, and Tectonics', *Science* **168**, 1084–1087.

Shor Jr, G. G., Kirk, H. K. and Menard, H. W.: 1968, 'Crustal Structure of the Melanesian Area', *Trans. A.G.U.* **49**, 217.

Sugimura, A. and Uyeda, S.: 1966, 'A Possible Anisotropy of the Upper Mantle Accounting for Deep Earthquake Faulting', *Tectonophysics* **5**, 25–33.

Sykes, L. R.: 1965, 'The Seismicity of the Arctic', *Bull. Seismol. Soc. Am.* **55**, 501–518.

Sykes, L. R.: 1967, 'Mechanism of Earthquakes and Nature of Faulting on the Mid-Oceanic Ridges', *J.G.R.* **72**, 2131–2153.

Sykes, L. R.: 1970, 'Seismicity of the Indian Ocean and a Possible Nascent Island Arc between Ceylon and Australia', *J.G.R.* **75**, 5041–5055.

Talwani, M., Pichon, X. Le and Ewing, M.: 1965, 'Crustal Structure of the Mid-Ocean Ridges. 2: Computed Model from Gravity and Seismic Refraction Data', *J.G.R.* **70**, 341–352.

Talwani, M., Pichon, X. Le and Heirtzler, J. R.: 1965, 'East Pacific Rise: the Magnetic Pattern and the Fracture Zones', *Science* **150**, 1109–1115.

Talwani, M., Windisch, C. and Langseth, M.: 1968, 'Recent Geophysical Studies on the Reykjanes Ridge', *Trans. A.G.U.* **49**, 201.

Talwani, M., Windisch, C. C. and Langseth Jr, M. G.: 1971, 'Reykjanes Crust: A Detailed Geophysical Study', *J.G.R.* **76**, 473–517.

Taylor, P. T., Brennan, J. A. and O'Neill, N. J.: 1971, 'Variable Sea-Floor Spreading off Baja California', *Nature* **229**, 396–399.

Tazieff, H.: 1971, 'Sur la tectonique de l'Afar Central', *C.R.* **272D**, 1055–1058.

Thellier, E.: 1966, 'Le champ magnétique terrestre fossile', *Nucleus* **7**.

Toksöz, M. N., Minear, J. W. and Julian, B. C.: 1971, 'Temperature Field and Geophysical Effects of a Downgoing Slab', *J.G.R.* **76**, 1113–1138.

Torrance, K. E. and Turcotte, D. L.: 1971, 'Structure of Convection Cells in the Mantle', *J.G.R.* **76**, 1154–1161.

Tozer, D. C.: 1967. 'Towards a Theory of Thermal Convection in the Mantle', in *The Earth's Mantle* (ed. by T. F. Gaskell), Academic Press, p. 325–353.

Vacquier, V.: 1965, 'Transcurrent Faulting in the Ocean Floor' in *A Symposium on Continental Drift* 77–81, *Ph. Tr.*, 1088.

Van Andel, T. H. and Bowin, C. O.: 1968, 'Mid-Atlantic Ridge Between 22° and 23° North Latitude and the Tectonics of Mid-Ocean Rises', *J.G.R.* **73**, 1279–1298.

Veevers, J. J., Jones, J. G. and Talent, J. A.: 1971, 'Indo-Australian Stratigraphy and the Configuration and Dispersal of Gondwanaland', *Nature* **229**, 383–388.

Verhoogen, E.: 1965, 'Phase Changes and Convection in the Earth's Mantle', in *A Symposium on Continental Drift*, 276–282, *Ph. Tr.*, 1088.

Verreault, F.: 1966, 'L'inversion des périodes propres de torsion de la Terre (3e partie)', *Ann. Géophys.* **22**, 131–146; thèse, Paris.

Vine, F. J.: 1966, 'Spreading of the Ocean Floor: New Evidence', *Science* **154**, 1405–1415.

Vine, F. J. and Matthews, D. H.: 1963, 'Magnetic Anomalies over Oceanic Ridges', *Nature* **199**, 947–949.

Vogt, P. R. and Ostenso, N. A.: 1967, 'Steady State Crustal Spreading', *Nature* **215**, 810–817.

Vogt, P. R. and Ostenso, N. A.: 1970, 'Magnetic and Gravity Profiles across the Alpha Cordillera and Their Relation to Arctic Sea-Floor Spreading', *J.G.R.* **75**, 4925–4937.

Vogt, P. R., Anderson, C. N., Bracey, D. R. and Schneider, E. D.: 1970, 'North Atlantic Magnetic Smooth Zones', *J.G.R.* **75**, 3955–3968.

Walcott, R. I.: 1970, 'Flexural Rigidity, Thickness, and Viscosity of the Lithosphere', *J.G.R.* **75**, 3941–3954.

Walker, G. P. L.: 1965, 'Evidence of Crustal Drift from Icelandic Geology', in *A Symposium on Continental Drift*, 199–204, *Ph. Tr.*, 1088.

Ward, P. L., Palmason, G. and Drake, C.: 1969, 'Microearthquake Survey and the Mid-Atlantic Ridge in Iceland', *J.G.R.* **74**, 665–684.

Weertman, J.: 1970, 'The Creep Strength of the Earth's Mantle', *Rev. Geophys. Space Phys.* **8**, 145–168.

Weertman, J.: 1971, 'Theory of Water-Filled Crevasses in Glaciers Applied to Vertical Magma Transport Beneath Oceanic Ridges', *J.G.R.* **76**, 1171–1183.

Whitmarsh, R. B.: 1968, 'Seismic Anisotropy of the Uppermost Mantle beneath Mid-Ocean Ridges', *Nature* **218**, 558–559.

Wilson, J. T.: 1965, 'Evidence from Ocean Islands Suggesting Movement in the Earth', in *A Symposium on Continental Drift*, 145–161 *Ph. Tr.*, 1088.

Wilson, J. T.: 1968, 'The Mountain-building Cycle and Its Application to the Cordillera', *Trans. A.G.U.* **49**, 327.

Worzel, J. L.: 1965, 'Deep Structure of Coastal Margins and Mid-Oceanic Ridges', in *Proc. XVIIth Symposium Colston Res. Soc.*, 335–361, Colston Paper XVII, Butterworths.

Wyllie, P. J.: 1971, 'Role of Water in Magma Generation and Initiation of Diapiric Reprise in the Mantle', *J.G.R.* **76**, 1328–1338.

GEOPHYSICS AND ASTROPHYSICS MONOGRAPHS

AN INTERNATIONAL SERIES OF FUNDAMENTAL TEXTBOOKS

Editor:

BILLY M. McCORMAC (Lockheed Palo Alto Research Laboratory)

Editorial Board:

R. GRANT ATHAY (High Altitude Observatory, Boulder)
P. J. COLEMAN, JR. (University of California, Los Angeles)
D. M. HUNTEN (Kitt Peak National Observatory, Tucson)
J. KLECZEK (Czechoslovak Academy of Sciences, Ondřejov)
R. LÜST (Institut für Extraterrestrische Physik, Garching-München)
R. E. MUNN (Meteorological Service of Canada, Toronto)
Z. ŠVESTKA (Fraunhofer Institute, Freiburg i. Br.)
G. WEILL (Institut d'Astrophysique, Paris)